Vision, Reflection, and Desire
IN WESTERN PAINTING

BETTIE ALLISON RAND *Lectures in Art History*

DAVID SUMMERS

Vision, Reflection, and Desire IN WESTERN PAINTING

THE UNIVERSITY OF NORTH CAROLINA PRESS Chapel Hill

Manufactured in the United States of America
Set in Minion type by Tseng Information Systems, Inc.

The publication of books in this series is made possible through the generous support of William G. Rand in memory of Bettie Allison Rand.

This book was published with the assistance of the William R. Kenan Jr. Fund of the University of North Carolina Press.

The paper in this book meets the guidelines for permanence and durability of the Committee on Production Guidelines for Book Longevity of the Council on Library Resources.

Library of Congress Cataloging-in-Publication Data
Summers, David, 1941–
Vision, reflection, and desire in western painting / David Summers.
p. cm. — (Bettie Allison Rand lectures in art history)
Includes bibliographical references and index.
ISBN 978-0-8078-3110-6 (cloth : alk. paper)
1. Optics and art—Europe. 2. Painting, European. 3. Visual perception. I. Title.
ND1158.O68S86 2007
759—dc22 2007011949

11 10 09 08 07 5 4 3 2 1

For Marilynne,

robins so red

CONTENTS

PREFACE

The lectures on which this book is based began some years ago with a telephone call from a longtime friend, Arthur Marks. We agreed on a series of Rand Lectures to be called "Light in Western Painting from Antiquity to Cubism." When writing actually began, the last lecture, to have been titled "The Painter of Modern Light," fell by the wayside as earlier art-historical issues demanded clarification and explanation. I am grateful to Arthur and his colleagues in Chapel Hill for having given me the occasion to write these lectures and to the Rand family for having made possible the distinguished series of lectures and publications to which I hope this book will prove a worthy contribution and thus a worthy memorial to Bettie Allison Rand. I am grateful to the colleagues and students in Charlottesville who listened to the first draft of the papers and of course to my second audiences in Chapel Hill, who heard the versions sharpened by that first presentation. I am grateful again to the Rand family and to Mary Sheriff and her colleagues in the History of Art Department at the University of North Carolina for their hospitality during my visit. At the University of North Carolina Press, Elaine Maisner and Paula Wald helped me through the ever-more intricate tasks of preparing a scholarly book for publication. I am grateful to my department chair, Lawrence O. Goedde III, and to the Lindner Family Trust at the University of Virginia for their contributions to the book's illustrations. Dwight Shurko's keen editorial eye was most helpful in the final preparation of the manuscript.

As always, I am thankful for the patience of my partner in life, my wife, Nancy, and for her remarkable ability to keep ongoing and clear track of an argument distant from her own concerns without breaking her own stride.

INTRODUCTION

In the opening pages of his *De architectura*, the Roman architect Vitruvius tells us that geometers, in addition to geometry, treat questions about vision, "which the Greeks call *logos optike*"—that is, the reasons of sight.[1] This *logos* was already some four hundred years old when Vitruvius wrote, and he assumed that its principles should be made part of the architect's professional judgment in order to determine how light will fall on a building, for example, or how the proportions of the orders should be adjusted to the circumstances in which a building is seen. He also described the adaptation of optics to architectural drawing and to illusionistic theater painting. This integration of Greek optics into the various practices of architecture is merely one episode in a long tradition stretching to the present, and in this book I will be concerned with ancient optics in relation to painting and with a later episode in the same tradition, the intersection of optics and painting in early modern Europe. The late Middle Ages and Renaissance witnessed the emergence of a new optics, together with great new traditions of painting. This period also coincided with European global expansion, a vast enterprise to which optics and its related disciplines and practices were by no means unrelated. By the mid–fourteenth century, foundations were being laid for the now-ubiquitous technological modeling and extension of sight, from microscopes and telescopes to television, with which characteristically modern lives are ever more thoroughly integrated.[2] The pervasive modern technological imitation of vision descends from much the same tradition of theory—as well as of experimentation, practice, invention, and implementation—of which Vitruvius wrote.

The variously applicable "reasons of sight" of Greek optics, as Vitruvius noted, consisted in large part of the *geometric* description of light in relation to the human organ of sight. The points, lines, and polygons of geometry are two-dimensional and are easily transformable into three-dimensional solids. Ancient and early modern writers referred to the "angles" and "pyramids" of vision, both with an apex at the eye, the angle with a line signifying a quantity at its base, and the pyramid with a shape signifying an area at the base. Vision need not be described in these culturally specific terms, of course: not only have innumerable human lives been conducted entirely without them, but—turning specifically to painting—virtual spaces of many kinds have been made without recourse to this assumption. When this geometric description of light (and space)—always in relation to the eye—began to be translated into *virtual* space, however, it gave rise to a new cluster of technical problems, demanding new skills of drawing, chief among them foreshortening. As these skills were perfected, they led to other optical issues (light, dark, and color) and to further questions about light, such as its interactions with atmosphere. Beyond such technical problems, geometric form and space in the virtual dimension engaged and enriched the temporality and, more specifically, narrativity inherent in virtual presentation itself—all the possibilities of apparent presence but also of past, future, and fantasy.[3]

Geometric optics is a representation of vision before it is a specifically pictorial representation of space and light. Both optics and the application of optical elements in painting—diagonal lines (foreshortened edges) and optical planes (foreshortened rectangles) as well as shadows, modeling, and highlights (the elements of what I shall call *optical naturalism*)—*entail point of view*. If from time to time all of these features are unified with respect to a "point of view"—in one or another system of "perspective" (from *perspectiva*, the late Latin translation of "optics")—absence of such unification should not conceal the fact that these optical features are part of a single system, and any one might always imply the others. In one or another combination, they provide the basis for the historical description of deep and perennial features of Western art. To be sure, instances and replications of these features show a wide range of thought, precision, or sophistication, and most are rooted in craft habit rather than contact with the textual tradition of optics. Still, as we shall see, the features of optical naturalism—diagonal lines and optical planes, for example—are persistent and *characteristic* in Western art, even in the art of the Middle Ages.[4]

Vision is sight and seeing, but the same word also refers to many kinds of mental activities, and although it is hard to explain how or why it is so, we know at once what it means to "see" a solution to a problem, to "visualize" a future event, to "remember" or "recollect" a past one, or to "reflect" on our lives. These higher-order "visions" are in fact ongoing, sometimes separable only with difficulty from our experience of the here and now. Furthermore, these higher-order visions, like sight itself, are shaped and directed by culturally specific spaces, times, and habits.

As the range in the meaning of "vision" might suggest, appeals to experiences to which the word might apply extend well beyond painting. In the Western classical tradition, rhetoric (the art of persuasion) was said from its beginnings to have succeeded in its aim—that is, to have changed minds and influenced actions—when it had met the standard of appearance. As Aristotle wrote, "one word is more proper than another, more of a likeness, and better suited to putting the matter before the eyes. . . . [I]t does make a difference, for instance, whether one says 'rosy-fingered morn,' rather than 'purple-fingered,' or, what is still worse, 'red-fingered.'"[5] Each of these still strikes our imagination very differently, disposing us toward both morning and the day to follow in very different ways, and it is not hard to see the possible dangers of such manipulation of imagination in a criminal trial, for example, or in political discourse, where it raises all the dangers of demagoguery. (The same dangers are hardly confined to the classical past, of course, and have vastly multiplied.)

Aristotle was duly suspicious of the art of rhetoric and its power over "sight" and action, but he also accepted it as a fact of life to be understood and managed. His teacher, Plato, as we shall see at length, vigorously resisted the rhetorical manipulation of "vision," part of his deep distrust of the sense of sight itself. His objections have had pervasive consequences in criticism to the present.

Poetry made a comparable address to vision (and was banished with the other imitative arts from Plato's republic). When Horace wrote the phrase "*ut pictura poesis* [as painting, so poetry]"—a phrase to be identified with the "humanistic theory of painting" in modern scholarship[6]—he echoed the much earlier Greek poet Simonides, who said that "painting is mute poetry, poetry is speaking painting." The common term shared by the two arts is visual, and Plutarch, who reported these words, compared the goal of writing to the painting of Apollodorus of Athens, "the first man to discover the art of mixing colors and of modeling of shadows."[7] I will discuss Apollodorus in detail in chapter 1. Still according to Plutarch, the narrative of the best historian is vivid like a painting, *enargeia* making the reader a spectator to the account.

Another book might make some headway toward understanding why sight, mind, science, and art came to be knotted together in just the way they were in the long European and broader Western tradition I shall be considering. It was argued early in this tradition that the disparate data of the five senses reported to an immediately higher, central entity—Plato's *psyche*, Aristotle's *koine aisthesis* (or common sense)—where sensation *as* sensation was united and represented to other mental faculties in potentially intelligible form.[8] These central, unifying faculties were closely related to memory (which retains these synthesized sensations) and imagination (which brings the absent to mind). They also provided a representation of the modes of sensation to reason. In this scheme, sight was preeminent among the senses because it conveyed most about the world to the mind. It conveyed lighted forms and therefore light and form.

The forms given to sight thus became an essential link between sense and mind. The word "idea" means "appearance" or "look," corresponding to the Latin *species*; but in Plato, "ideas" are purely mental, visible only to the eye of reason. For Aristotle "form" is manifest only in real things but is also abstractable from them and so may assume teleological significance as formal and final cause. Even if unmoored from their ancient philosophical beginnings, these formulations have persisted to the present. The same assumptions raised other possibilities. Sight may also be deceived. The Greek *skopein* (to look at) has given us "telescope" and "microscope" as well as "skepticism," which might be supposed first to have meant to look again, not to accept what is first seen. This places the burden for validation not only on sense itself but also on memory and experience. And of course a more radical skepticism could rebound through the whole vision-based system. The sophist "Gorgias maintains first, that nothing exists; secondly, that if anything exists it is unknowable; and thirdly, that if anything exists and is knowable, it cannot be demonstrated to others."[9]

If our apprehension of the external world is the representation of present sensation to the mind, then the world is always apprehended at a remove, and this entails a fundamental and characteristic project for imitative art. Not only are appearances to be imitated, but they must be given the force of presence by artificial means. This leads us back to the close kinship of painting, poetry, and rhetoric, which may again be seen to pursue the same end. *Phantasia*, which first meant "presence in the light," became the term for the capacity of the mind to bring the absent into the mind's light, before the higher vision of the mind's eye. The absent—the past, the future, the imaginary—should be made as if present by art. In arguments that were to clear the way for centuries of Western critical thought, Plato compared rhetoric to the painter's imitation of appearances (as opposed to one or another kind of mimicry or to the imitation of state of mind or feeling by music). The art and skill of the painter make us see what is not there, see three dimensions in two, and it is not surprising that, as I will argue, the beginning of the imitation of appearances in Greek painting and of the traditions of imitation stemming from Greek painting to the present were closely linked to the beginnings of the study of optics.

As James Ackerman observed, optics was one of the most highly developed fields in antiquity and the Middle Ages because it could be pursued to considerable lengths with existing mathematics and because it contributed to fundamental epistemological problems.[10] In fact, optics has had a nearly continuous history from the Greeks to Islam, then back to Europe to the present. Optics was also an early experimental science, and the word "demonstration" appears fairly often in the Renaissance texts we shall be examining. The proof of a demonstration, however, might consist of being fooled by an illusion, as we shall see Filarete describing *disegno* itself as a demonstration of that which is not.[11]

The fundamental analytic device of ancient, medieval, and Renaissance optics was the *visual angle*, a triangle with its apex at the eye, its base at the object seen, and the two connected by straight lines according to the economy of light. This formulation of vision, which seems simple at first blush, actually raised the deepest questions about the relations among sense, mind, and the world, questions that have persisted to the present. What are color and light? Why does light behave with such economy? How does the geometry of light relate to the physiology of the eye (and eyes, although classical and medieval optics was almost entirely monocular) and brain? Much of the most fundamental natural science descending from the ancient Greeks has in fact been concerned with these questions.

The problem of the nature of the "lines" involved in the "geometry" of sight was in play from the start. From the first mentions in ancient texts, optics was regarded as what medieval Scholastics called a "middle science," one that applied geometric arguments to the explanation of natural phenomena. In these terms, geometry itself was pure and speculative, its points and lines immaterial, in principle existing only in thought. Optics was relatively *impure*, applying relevant geometric principles to physical rays of light, just as, Aristotle wrote in his *Physics*, other physical sciences like astronomy might apply geometric arguments to planetary orbits or harmonics might apply such arguments to problems of acoustics.[12]

To make a long history short, the "middle sciences" were developed apace in the Islamic world, then in the later Western Middle Ages, and the principal writers I will discuss moved forward from these advances.[13] During the same centuries, the "middle sciences" also came to be firmly linked to the mechanical arts. Although Leon Battista Alberti (figure I.1), the main character in the following chapters, whose *De pictura* was the first modern treatise on the art of painting, appealed to the "philosophers" from time to time, he also insisted that he was writing as a painter and for painters and that in formulating his new art from the principles of nature he would pursue "*una più grassa Minerva* [a plumper Wisdom]."[14] Alberti's reference is to Cicero,[15] but I understand it to be a humanist's way of saying that he was altogether aware of writing for practitioners of a middle science. Moreover, Alberti dedicated the Italian version of his text not to a painter but to a master of other, related middle sciences, the architect Filippo Brunelleschi.

If optics presumed its lines to be physical, this by no means defined their nature. Alberti began his treatise in a Euclidean manner, with the elements of geometry, but immediately defined these elements in extra-Euclidean terms.

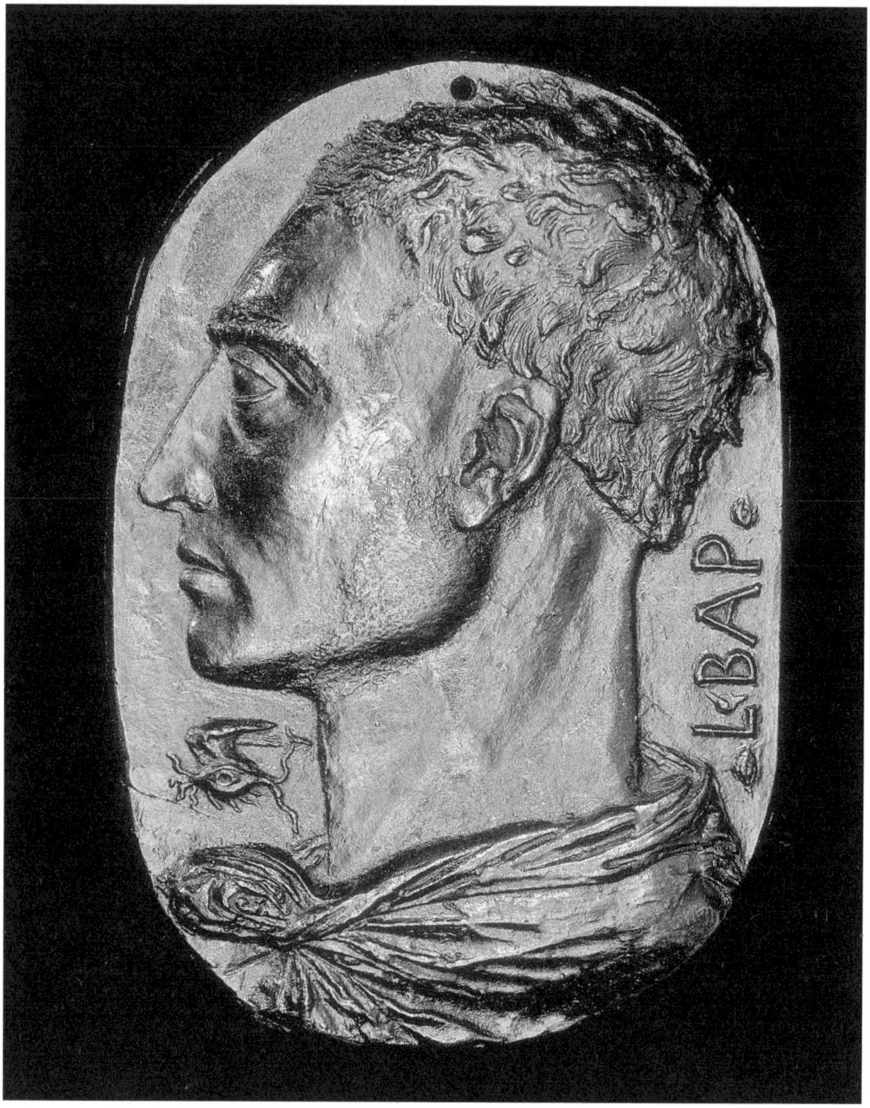

FIGURE I.1. *Leon Battista Alberti,* Self-Portrait, *bronze, ca. 1435, National Gallery of Art, Washington, D.C. Samuel H. Kress Collection, © 2006 Board of Trustees, National Gallery of Art, Washington, D.C.*

A point is a least visible "sign" on a surface, and its analogue (or metaphor) in painting is the smallest mark a painter can make. Surfaces made up of such points constitute the visible, the true subject of the painter. A surface is made up of lines "like the threads in a cloth." When Alberti wrote that a line is "so slender in width that it cannot be split,"[16] that fine line stands for a colored and lighted surface sliding out of the viewer's sight. If Alberti was writing to raise the status of the art of painting, he was also writing to raise the status of the mechanical arts, of applied geometry and mathematics, of arts producing effects the causes of which could not be understood, like Brunelleschi's marvelous dome of Florence Cathedral, and the irresistible illusions of painting.[17]

Alberti's friend, Cristoforo Landino, writing in the prooemium to his commentary on Dante's *Divina commedia*, first published in 1482, included Alberti among the Florentines excellent in learning.[18] First, Landino placed Alberti among the physicists: "he was born only to investigate the secrets of nature." But then he had mastered the whole quadrivium (geometry, arithmetic, astronomy, music) and the new addition, optics (*perspectiva*), in which "he was marvelous beyond anyone for many centuries."[19] He wrote eloquently on architecture, on painting, and on sculpture. "Not only did he write, but he made works with brush, chisel, burin, and bronze." Although no paintings have been convincingly attributed to him, Alberti insisted on the value of his experience as a painter, and his historical importance as an architect is well known.[20] Giorgio Vasari, writing in the middle of the sixteenth century, took a dimmer view than Landino of Alberti's talent, which Vasari considered to be undisciplined by *disegno*,[21] as in fact the aristocrat-humanist Alberti had not been trained in a shop from apprentice to master. A pen-and-ink self-portrait attributed to him is admittedly unprepossessing, but then it is early in the history of Florentine drawing taken altogether, and the works in sculpture attributed to him, including the self-portrait in figure I.1, are more impressive. As a painter, Alberti may be supposed to have been an amateur in the fullest sense of the word, a lover of the art, although it is fundamentally important that his writings located the art in the broader context of the middle sciences. And, if the sample is small, it is perhaps also significant that three of the four works arguably from his hand are self-portraits.

Landino's addition of optics to the quadrivium was a culmination of the rise and diffusion of the new science in the late Middle Ages, especially among writers belonging to the Franciscan order. Another late-Quattrocento testimonial to this change is to be seen in Antonio del Pollaiuolo's bronze tomb of the Franciscan pope Sixtus IV (figure I.2). The smiling personification of "Prospectiva" reclines beside a flame, the implied shadow of which must be meant

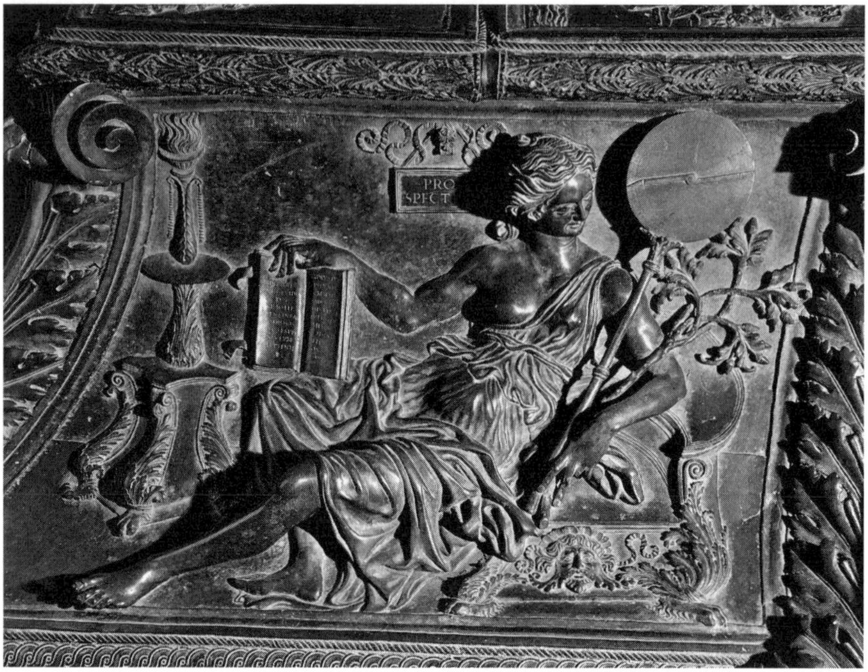

FIGURE I.2. *Antonio del Pollaiuolo,* Tomb of Sixtus IV, *completed 1493, detail,* Personification of Optics (Perspectiva), *Museum of the Treasury, St. Peter's, Vatican. Scala/Art Resource, New York.*

to illustrate the first principle of optics—that light travels in straight lines.[22] Principles of direct, refracted, and reflected vision from the thirteenth-century *Perspectiva communis* of John Pecham are inscribed on the pages of her book, and, as Samuel Y. Edgerton Jr. noted, she holds an astrolabe in the crook of her left arm, thus expanding optics to the scale of cartography, cosmology, and astronomy.[23] In her excellent study of Alberti, Joan Kelly observed that he used the same "horizon," a circular disk with divisions, for his *Descriptio urbis Romae* (map of Rome) and for his *De statua* (treatise on sculpture).[24] It seems unlikely that the delicate system of pointing Alberti recommended to the sculptor could have been very useful, but the system of measurement itself places the standing human figure at the center of its visible and measurable world, a world in this case coincident with the city of Rome viewed from the Capitoline. Had Alberti's "horizon" been placed on edge, it would have been an astrolabe, thus placing the proportional human figure at the center of its moving horizon, encompassing Rome, beneath the legible order of the heavens.

Alberti's great successor, Leonardo da Vinci, wrote in a grand summary of his *paragone* arguments that the eye "has measured the heights and size of the stars . . . found the elements and their locations; it counsels and corrects all human arts, it moves men to different parts of the world. . . . [I]t has generated architecture, perspective and divine painting."[25] Both Leonardo and Alberti practiced astronomy, cosmology, cartography, architecture, and painting to one or another extent, and all are subalternate geometries that could be united in Leonardo's mind by what he called "visual lines."[26] Moreover, these geometries of visual lines, again inseparable from mechanical arts, were part of what I will call a *specialist culture* of disciplines with more or less common and mutually explainable practices and geometric principles. That is, all of these "middle" physical geometries, or applications of physical geometries, could be worked out as descriptions of interactions of thought, imagination, sight, and visible reality—in drawings, which are "optical" in the broad sense in which I am using the word. It is not a coincidence that the art of drawing, as a means both of analysis and invention, also came to its first full flowering in the hands of Leonardo da Vinci.[27]

Alberti, who died when Leonardo was in his early twenties, will figure in all of the chapters to follow. Alberti drew on the same sources I will consider in chapter 1, and it was Alberti who called the eye a "*speculum animatum*," a living mirror, the title of chapter 2. Alberti was engaged in the cartography I will discuss in chapter 3, and he liked to tell his friends that the poets were right, that Narcissus was the inventor of painting, providing half the title for chapter 4.

SPECIALIZATION

If geometric optics itself is a representation of vision, then it must also be a schema for the imitation of appearances. Optics, however, was only partly concerned with diagramming the activity of light in vision; it was also a highly developed intellectual historical tradition with many connections and tributaries. What I shall call the *principle of specialization* means that anyone undertaking the systematic representation of appearances, rather than simply looking at things or picking solutions equivalent to centuries of thought and experiment out of the air, would turn to this specialized discourse, with all of its possible associations. It is clear that the principles of the new optics were available in Florence by 1400, no doubt in response to prevailing local interest, and it is just as clear that some painters wanted to adapt these principles to the purposes of their art. There must have been other adaptations outside the art and craft of painting. As the understanding of optical principles became more common, for example, it must have become apparent to some that the apex of the visual pyramid could be refracted with a bit of glass (a lens, Latin for "lentil"), thus to correct vision.[28] It can hardly be a coincidence that the appearance of eyeglasses coincided with the diffusion of optics.

SPECIES, EXTRAMISSION, AND INTROMISSION

The notion persisted in the optical tradition that there is a generative image relation between external things and the "images" we have of them in our eye and mind. In atomist theories of vision (Democritus, Epicurus, and Lucretius), bodies were presumed to issue eidola or simulacra (images of themselves that impressed the material medium of air) and thus to pass to and impress the eye. In the Middle Ages, the related idea of *species*—which most simply means "appearance" and corresponds to the Greek *idea*—underwent a complex conceptual development and lived on well into the Renaissance.

Given the visual angle, do the rays forming it go out from the eye (*extramission*), or do they come into it (*intromission*)? In very general terms, extramission is ancient, more compatible with notions like *species*, and intromission is modern, more compatible with the punctiform description of light. Plato came to be closely associated with extramission, although to the degree that he had a fully developed explanation of vision, he defended a subtler scheme according to which in daylight light from objects coalesces with "light" issuing from the eye in the act of vision. The compendious and authoritative Euclid also assumed extramission, and Ptolemy followed.[29] As I have said, versions of extramission were repeated into the Renaissance.[30] Piero della Francesca, to take a significant example, described an extramissive scheme of vision as elegant as his perspective diagrams. The eye, he said, is "spherical [*tondo*], and from the intersection of two little nerves which cross one another the power of sight [*virtù visiva*] comes to the center of the crystalline humor, and from that the rays set out and extend in straight lines, passing through one quarter of the circle of the eye; so that this part [the angle of vision] subtends a right angle at the center."[31] We will encounter this sequence in variations in the chapters to follow, usually, however, presented less geometrically and running in the opposite direction—that is, intromissively—with rays coming into the eyes.

Aristotle is closely associated with the beginnings of intromission,[32] with the idea that the eye is passive in sight or that it is active only in the sense of being uniquely susceptible to color in light. Here it will be important to distinguish carefully between two analogies of sight to touch. Extramission is comparable to touching (or grasping) external objects; intromission is like *being touched* by light from them. Aristotle corrected but did not reject outright the argument of the atomist Democritus that the act of vision is the reflection of eidola emitted by objects on the cornea of the eye, which Democritus called "emphasis."

If sight is that simple, Aristotle objected, then the physiological complexity of the eye cannot be explained. The act of sight *is* in fact like a reflection, Aristotle wrote, and Democritus was right to think of the eye as watery, but the act of vision, the "reflection," must occur inside it.[33] As we shall see, Galen later located the act of vision on the crystalline humor, and in the early seventeenth century, Johannes Kepler, redefining the crystalline humor as a lens, finally formulated the modern retinal (and intromissive) theory of vision.

Although extramissive formulations persisted, intromission was given the foundations necessary for its triumph in the great optics of the Arab writer Ibn al-Haytham, called Alhazen by his Latin translators. Alhazen, who wrote at the turn of the tenth and eleventh centuries and was translated into Latin at the beginning of the thirteenth, established the principles on which the more or less continuous tradition of late medieval and modern optics would be based.[34] Alhazen's optics were the first based on an adequate notion of the propagation of light, which made it possible to describe any act of vision as the interaction of surfaces—that is, the transfer of a set of points from external surfaces to the surface of the eye. This in turn made it possible to eliminate the idea of species once and for all. For Alhazen, the eye (and Galen's crystalline humor) was a surface reacting among surfaces, and from these reactions higher mental faculties made inferences and judgments. More than simply a technical term, "intromission" has been of fundamental importance for the understanding of the relation of mind and world, at least on a par with the understanding of visual "lines" as physical.

ARISTOTELIAN COLOR THEORY

I will refer fairly often to Aristotle's account of color, an always available adjunct to the optical tradition. This "theory" of color, which is scattered in Aristotle's own writings, was gathered, outlined, and expanded by his many followers and commentators. The basic scheme was sufficiently serviceable to painters to have lasted well into the nineteenth century—that is, well beyond Newton's spectral analysis of white light.[35] According to Aristotle, and in the ancient tradition generally, colors (*chromata*; in Latin, *superficies*, both meaning "surface")[36] are qualities of objects perceivable by sight in the presence of a lighted medium, not subjective reactions to frequencies of light, as we have come to understand them. For Aristotle, all visible colors were a mean between light and dark in the sense that the organ of sight is not suited to the apprehension of extremes; the eye does not function if there is no light or if there is too much light.[37] Colors (or hues) themselves were considered mixtures of white and black and were arranged according to what we now call *value*—that is, relative light and dark. Aristotle himself was not rigorous in defining this range, but in general blue is closer to black, yellow closer to white, red and green somewhere in the middle. He also considered the possibilities that the mixtures of light and dark constituting colors might be able to be defined in terms of proportions comparable to the harmonies of music or that they might be formed by superimposition, as when painters lay a color over a brighter one.[38]

The discussion of color in Aristotle's account of rainbows in the *Meteorology* cast a millennial shadow over speculation on the nature of color. Complementing his other discussions, it stated that colors appear to be different in different light and in relation to other colors. The colors of the rainbow, Aristotle said, are most intense against the darkest clouds, and yellow arises from the contrast of red and green. Aristotle also associated black with the weakening of sight over distance. He might have meant that distant forms seem dark against a bright horizon, as they sometimes do, but more generally he seems to have meant that things become "darker" as they become less distinct. This is an explanation of "diminution" in terms of light but not in terms of the geometry of light.[39] A late classical author, Johannes Philoponus, the first Christian Aristotelian, writing in a commentary on the *Meteorology*, related the equation of near and white and of dark and far specifically to the practice of painters. According to what E. H. Gombrich dubbed the "rule of Philoponus," if white and black are placed on the same surface, white will appear to be nearer, black to be farther away, so that when painters paint a "cistern, ditch or cave," they paint it black or dark brown, and they use black around forms meant to look prominent "such as the breasts of a girl, an outstretched hand, or the legs of a horse."[40] Philoponus thus coupled white with foreshortening—"an outstretched hand"—and in general, this kind of modeling showed forms with white at the center, turning to black at both sides, as if the viewer's eye (or painter's eye) were the source of light. As Gombrich points out, the "rule of Philoponus" was still followed in shop practice in the Renaissance, and, more immediately in the tradition of Aristotle's own texts, Leonardo in his full maturity grappled with the implications of the equation of darkness and distance.

MEASURE AND RATIO

Alberti set forth his famous metaphor of painting as an "open window," together with his one-point modular construction, near the end of the first book of his treatise on painting. What was to be seen through this window? For a viewer, it was most deeply a world of relations, of implicit quantities and qualities, but of *apparent* quantities and qualities. Alberti cited "the philosophers," who say that if the whole universe and everything in it were to be diminished by half, there would be no evident difference because all relations among things would still be the same.[41] From this he drew a general principle regarding both quantities and qualities *as they are apparent to us*. It might thus be better to ask what is *on* Alberti's window because, as we shall see, it is on the window, the plane of intersection, that actual ratios (not real measures) are to be seen. Alberti wrote that

> large, small; long, short; high, low; broad, narrow; bright, dark; luminous, shadowy, which because they may or may not be joined to things, are called "accidents" [as opposed to substances] . . . are such that any knowledge of them is only possible through comparison. White ivory and silver placed next to a swan or snow seem merely pale. For this reason in painting things appear most splendid when there is a good proportion of white and black. . . . [C]omparison . . . immediately shows in a thing what is more, less, or equal. . . . [W]e compare things to what is best known, and since man is the best known to us, perhaps Protagoras, saying that man was the mode and measure of all things, meant that all the accidents of things are known in comparison to the accidents of man.

What are the "accidents of man"? Perhaps Alberti meant no more than that all standards of measure—hands, feet, *braccia*—are convenient but arbitrary and therefore are most valid as relation, like the modular *braccia* governing his construction. But of course the list of the "accidents of man" in relation to which we know the world may be considerably expanded. Alberti's extraordinary and original transformation of painting into a metaphor for our subjective experience of the world taken altogether places him alongside Nicholas Cusanus as one of the first to articulate a principle of emerging modernity.[42] And when Alberti appealed to the philosophers, it was the archskeptic Protagoras that he finally cited. By doing so, Alberti opened the way for those human capacities that make appropriate judgments among particular, circumstantial, and relative things. The world in which we live—or a fictive world like the one in which we live—cannot be truly represented otherwise than as the continual occasion and demand for such judgments.

At the beginning of book 3 of this treatise, Alberti wrote that the task of the painter is "to describe with lines, and tint with colors, on whatever panel or wall is given to him, the surfaces, like those seen, of whatever body, so that those [surfaces] at a certain distance and at a certain central position seem raised up [*rilevate*] and very similar to having bodies." Grace, goodwill, and praise,

Alberti continued, will come to the painter whose painting "will hold the eye and mind of whoever sees it; we have explained how to do this above when we treated composition and the reception of light."[43]

As Michael Baxandall showed, Alberti's pictorial "composition" was adapted from classical rhetoric,[44] and as I have said, classical rhetoric itself is deeply visual in that it achieves its end of persuasion when the skillful use of words has placed the matter as if "before the eyes" of the audience. Clarity of organization and vivid description contribute to this address to sight, but composition also raises more complex issues of ornament; of the use (and abuse) of figures, tropes, and "colors"; and of ingenuity in general—all those artful devices that make discourse "brilliant," pleasing, and therefore persuasive. (Composition is "persuasive" and not "convincing" because the primary appeal of rhetoric is to seeming and imagination, not to logic and reason.) Alberti's references to light and relief as well as to distance and position, however, raise the extracompositional issues—and the specifically optical issues—of what we call "perspective." I will be concerned with perspective in these chapters, but I will also return several times to the following phrase. Alberti said that painters have achieved their goal "when their painting *will hold the eyes and mind* of whoever sees them" (in the Italian text) or (in the Latin) when the painting "will *hold the eyes and move the mind* of the viewer."[45] In these two versions, "holding the mind" and "moving the mind" of the viewer point once again to the example of rhetoric, which might persuade (but again, not convince) by "moving" listeners. The phrase "hold the eyes," however, is common to both versions.

When Alberti thus summed up the principles of his new art of painting, he echoed the words of Pliny the Elder, who wrote in the history of painting in his *Natural History* that before Apollodorus of Athens, who was called Skiagraphos (Shadow Painter) and who (as we have seen in our discussion of poetry) first modeled his figures, there were no paintings to be seen that "might hold the eye."[46] For Alberti, the reception of light (*rilievo*) and composition—two kinds of brilliance—effected the first visual fascination.

Finally, the question of the interrelation of optics and painting raises the more general question of the use of simple optical devices by painters in the longer Western tradition, which historians of art are mostly inclined to avoid or reject outright in favor of what is implicitly a simple notion of art as expression.[47] I will argue that painters at crucial points learned (or were taught) to see not so much through passive reliance on mirrors or *camere obscure* but rather through taking part in the broadly emerging consensus that vision itself is comparable to what is to be seen on such surfaces and that it is therefore possible to learn about appearances through them. If all aspiring painters were given a *camera obscura* or a *camera lucida*, few would come away painting like Vermeer or drawing like Ingres (which is not to say that some might not make very good paintings or drawings). Before the triumph of the modern idea of art as expression of personal vision, painters might be supposed to have willingly and unself-consciously incorporated what was to be gathered from the aids available to them in the pursuit of the fascinating vividness of which Pliny and Alberti wrote.

CHAPTER 1 Shadow-Painting and Scene-Painting

The Beginnings of Optics and Classical Art

In this chapter, I will argue that Greek painting and the art (or as we now call it, the science) of optics started out together. We are very inclined to think that appearances are simply given to us, in which case their imitation requires no more than close observation. As is well known, however, E. H. Gombrich argued instead that there is no such "innocent eye" and that in traditions in which the imitation of appearances becomes an issue at all, appearances can be addressed only from the historical standpoint of existing images and skills.[1] I will extend Gombrich's position to argue that the desire to describe appearances evident in Greek painting cannot be separated from the desire to analyze and understand appearances themselves in more general terms.

To be sure, Greek art displays resolute progress in observation and descriptive skills, but it also displays two fundamental abstractions from the bright complexities of vision, principles that, however commonplace they may seem, cannot be supposed to have resulted from simple looking, no matter how careful.[2] These two principles—the definition of the visible in terms of light, dark, and color and the geometric economy of light—are shared by later fifth-century Greek painting and optics, and neither the unique importance assumed by light in European painting nor the dynamics of European visuality and its dependencies to the present can be understood without taking into account the interrelation of the pictorial and the explicitly optical.

Aristotle wrote—some time after the first painters I shall be discussing—that the special object of vision is color, apprehended under varying conditions of light and darkness,[3] a simple analytic formulation that takes us to the first of the principal topics of this chapter, the modeling of virtual forms, which the Greeks called *skiagraphia* (shadow-painting). Aristotle also wrote that the transmission of color in light requires a lighted transparent medium and that colors are not visible when objects are in actual contact with the eye. Vision, in other words, entails distance, through which light and color travel in straight lines, so that relations between visible things and the eye can be described in terms of geometry. More specifically, as I have described in the introduction, the activity of light in relation to the organ of vision came to be treated by Greek optics primarily in terms of the *visual angle*, the calipers of sight with its apex at the eye, its base at the object. This fundamental tool of optical analysis will take us to the second principal topic of this chapter, to ancient Greek *skenographia* (scene-painting), its name preserving its origin in theater.

My arguments begin along the path marked by Gombrich in an article, "The Heritage of Apelles," published in 1976.[4] Apelles was the legendary greatest painter of antiquity, the designated painter of Alexander the Great. No paintings by Apelles survive, although it has been argued that the *Herakles Discovering Telephus* from the basilica at Herculaneum (figure 1.1) is a close and well-preserved copy of one of his works mentioned by Pliny the Elder. The painting from which the mosaic *Battle of Alexander the Great and Darius* (figure 1.12) was made has also been attributed to Apelles.[5] Gombrich argued that Apelles contributed a crowning illusionistic device—the highlight—to Greek painting, a culminating development that, as we shall shortly see, began with modeling and (perhaps) with cast shadows. In figure 1.1, the seated figure of Demeter, with a splendid still life of grapes beside her and a piping Pan behind, faces fully into the light, which falls from our right. Shadows are subject to the economy of light and thus clarify regions of virtual space. The swarthier male hero casts a shadow toward Demeter, and light is reflected to us in a highlight on his glistening body from the same source. Versions and variants of this cluster of optical features—modeling, highlights, shadows, diagonal lines, and optical planes—has persisted in painting from the later fourth century B.C.E. to the present and, as Gombrich also argued, spread beyond the borders of the world known when Apelles painted.

At the beginning of his article, Gombrich observed that while art historians continue to lavish attention on the representation of space, they have largely neglected what the poet Goethe called "the action and passion of light." In fact, although one or the other may be emphasized, represented space and light are not opposed, as Gombrich's words might be taken to suggest; on the contrary, they entail one another. What we call "perspective" is nothing but a schematic geometry of light, part of the regular and predictable activity of light but, in relation to a single point, an equally schematic eye.

Few readers will have heard of Apollodorus of Athens (said to be the first to model his figures) or Agatharcus of Samos (the first to paint the illusionistic architecture of *skenographia*). No one seems to doubt their flourishings, however, and we might put a little imaginative flesh on their scant historical bones by surmising that they must have known one another since they shared related interests in Athens in the mid- to late fifth century B.C.E.[6]

The innovations of these two painters had consequences I will be gathering throughout this book. In explicitly acknowledging the physical conditions of sight—color, light, and dark—Apollodorus also acknowledged the *individual* experience of sight. For his part, Agatharcus first constructed the geometry of light as a spatial relation to a viewer—*any* viewer but also, at least implicitly, always an individual viewer. And, to restate my premise, these pictorial inventions, rather than simply being "optical" in the sense of recording what we see or in the sense of being "influenced" by optics, took place together with the beginnings of the emerging science of optics and in fact contributed substantially to its rise. To be sure, painting and optics were to go their separate ways, and the later interactions of the two are much more complex. But in the longer tradition, Apollodorus and Agatharcus (and later Apelles) began the history of the representation of the world from and for a point of view.

The devices of optical naturalism to be seen in *Herakles Discovering Telephus* (figure 1.1) are still closely associated with naturalism in general. It requires only a moment's thought, however, to realize that traditions of naturalistic painting have been highly developed in other terms. Neither Egyptian nor earlier Greek painting, which show individual persons with descriptive proportions engaged in real (if sometimes mythical) activities, show depicted light. Moreover, to take an example from a tradition historically unrelated to the Western tradition, Mayan painting (figure 1.2) shows what seem to be individual persons with similarly descriptive proportions; their forms are foreshortened as they seem to move freely, but the surfaces of their bodies are again uniformly colored and seem to exist in a kind of absolute surrounding daylight. The virtual space shows familiar forms of Mayan architecture, but these forms are not seen under a visual angle. In the terms I will be using, in short, Egyptian, early Greek, and Mayan painting are naturalistic, but they are not optical—or, perhaps more exactly, represented surfaces display only one optical element, color, which is not treated in relation to virtual light. To take another major but distant example, early Chinese painting is descriptive and still more three-dimensionally fluid, but there is no modeling. In Indian painting—in the frescoes at Ajanta, for example—figures are modeled with a kind of powdery delicacy,[7] but this, however characteristically it has been adapted, may be an echo of Greek painting. We may thus return to one of Gombrich's conclusions—that the activity of light in space and its interaction with forms is nowhere the positive, perennial theme it is in Western painting—and these brief comparisons must only raise more pointedly the question of how and why these deep optical themes appeared in Western painting and why they have both persisted and changed in the ways they have.

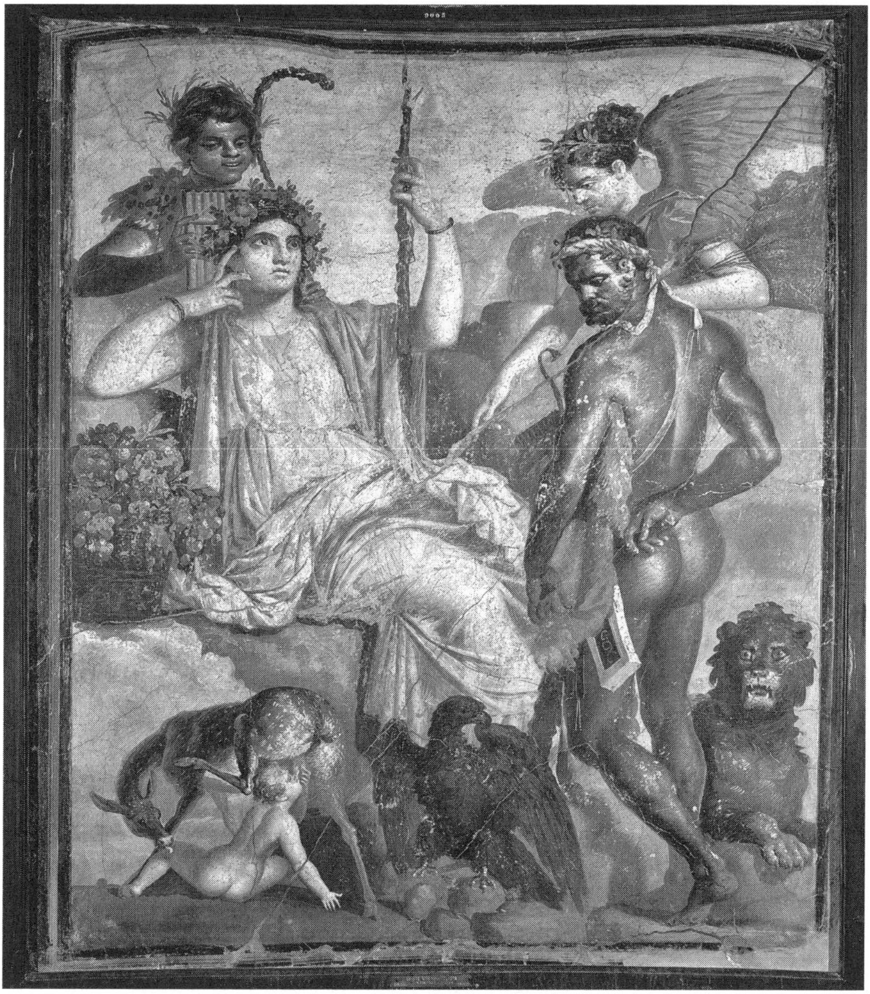

FIGURE 1.1. Herakles Discovering Telephus, *from the Basilica at Herculaneum, Roman copy after Greek original of the fourth century* B.C.E., *Museo Archeologico Nazionale, Naples. Scala/Art Resource, New York. (Reproduced in color as plate 1.)*

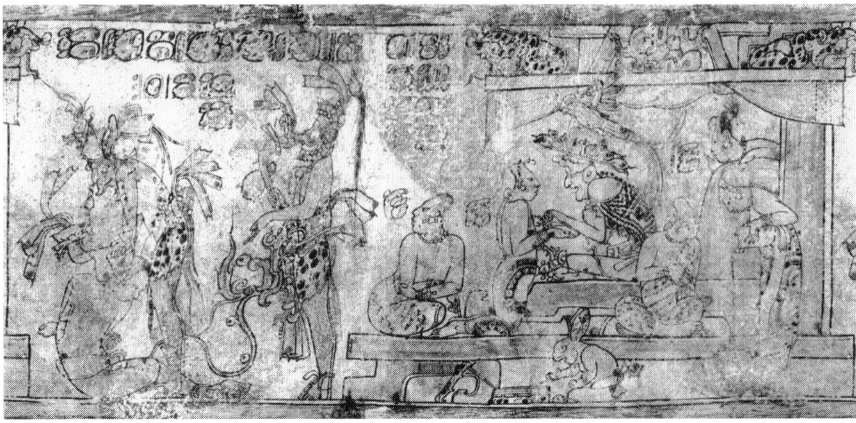

FIGURE 1.2. *Rollout of a cylindrical vessel showing a Lord of the Underworld with five goddesses and a sacrifice, painted ceramic on cream slip, late classic Maya, 600–900 C.E., Art Museum, Princeton University. Photo: Justin Kerr.*

Some metaphors for the characterization of light appeared long before the fifth century B.C.E. and have lasted to the present. We still speak of "rays" of light and of "radiation" of all kinds. The root of both words is *radius* (rod). Figure 1.3 is a representation (but not a depiction) of active light. Akhenaten and Nefertiti are in relief and so are modeled by the light we share with them; bars reach down from the similarly modeled disk of the sun, bearing beneficent energy, extending life and health. If the light of the sun is beneficent, however, the deeper characterization is that this beneficent energy (and light in general) works directly, following the least path—that is, moving in straight lines from its source to its recipient.

Lines are elements of geometry, and in the early sixth century B.C.E., the military engineer Thales of Miletus, semilegendary founder of Greek natural science and philosophy, who is said to have changed Egyptian land measurement (which must have involved surveying) to Greek geometry, is also said to have determined the height of the Great Pyramid at Giza by measuring the length of its shadow in relation to his own height at the same time, then computing the ratios of the two heights to the lengths of the respective shadows.[8] This basic trigonometric operation presupposes that height, length of shadow, and hypotenuse form a right triangle, which is subject to more general geometric rules, and the hypotenuse of these triangles would have been defined by the path—the straight line—of the pyramid's shadow. The clinching general principle is that similar triangles are proportional. Closer to the visual angle of Euclidean optics is the story that Thales devised a way of estimating the distance

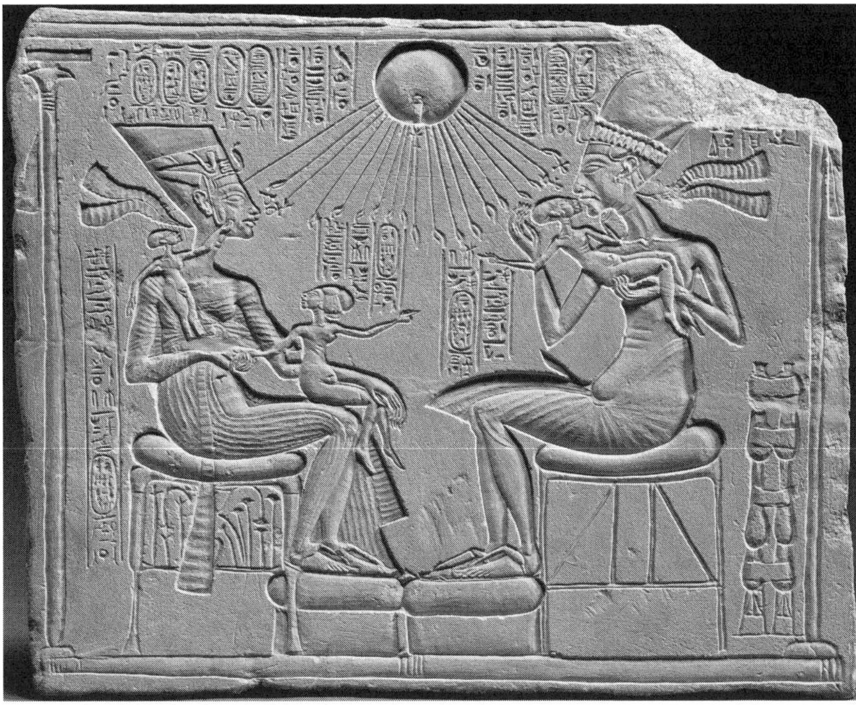

FIGURE 1.3. Akhenaten and His Family Worshipping the Sun Disk, *ca. 1345 B.C.E., Ägyptisches Museum, Staatliche Museen zu Berlin. Bildarchiv Preussischer Kulturbesitz/ Art Resource, New York.*

of boats from shore by their apparent size. If we know the height of a boat, we may align an upright between it and a point of sight, and by figuring out the ratios between the apparent height on the upright and the known dimension, we may judge the ship's distance. (This method of measurement was no doubt too handy not to persist, and some two millennia later, the thirteenth-century architect Villard de Honnecourt showed towers being measured in much the way attributed to Thales.)[9] For optical purposes, the crucial new connection was that between the geometry of light and the eye.

The foundational myth of European painting is about light and shadow. Pliny the Elder did not feel able to specify the dates and times of the invention of the art of painting, but he considered it agreed that the art began with tracing an outline around a man's shadow and that these paintings were monochromes, painted a single color. Ecphantus of Corinth, Pliny said, was the first to use color, using a pigment made of powdered earthenware, and early painters specified their paintings by writing on them the names of those represented.[10]

Pliny told the famous story of the Corinthian maiden who traced the shadow of her soon-to-depart lover at the beginning of a brief account of the "plastic" art appended to his history of painting. She had made a kind of assisted photograph—a light drawing—indexical in that the contour was determined by the fall of light but drawn and fixed by her hand. This outline was then filled in with clay by her father, Soutades, to become a relief plaque.[11]

We still speak of both sculpture and painting in terms of "modeling," and, as we have already seen in the introduction, Renaissance painters routinely referred to "relief" in painting, as if painting and sculpture were interchangeable. If Soutades had *painted* the equivalent of his relief on a panel, by modulating dark pigment inward from each side of the contour of a shape toward the middle, as if in response to a source of light, he would have made a simple shadow-painting. He would have made a two-dimensional shape into a virtual three-dimensional form, and this virtual "relief" would have established the equivalence of sight and touch—the presentation of the visible as if it could be touched—raising issues best known in the Renaissance *paragone* between painting and sculpture.[12] Finally, however, this ambivalence of sight and touch is rooted in the deepest assumption of optics—that it describes physical contact and process.

Skiagraphia and *skenographia* have been confused by translators and commentators since antiquity. Their common term is illusion, and, as we have seen, both belong to the same system. Since they involve different technical procedures, however, they must be carefully distinguished. Since *skenographia* displays more of the features of optical naturalism, I will begin with it, then turn to *skiagraphia*, before turning finally to reflection, which raises issues of the geometry of light more comparable to *skenographia*.

The advent of illusionism in painting and the beginnings of optics are deeply implicated in the history of theater. Aristotle wrote in his *Poetics* that song, diction, and the artful appeal to vision are the imitative means of tragedy.[13] That is, the imitative means of tragedy address *by means of art* the senses of sight and hearing. *Opsis* (often translated as "spectacle" but with simpler meanings of something seen) is—still according to Aristotle—one of the six parts of tragedy.[14] It may lead the mind in its own right, but it is not properly part of the art of poetry. It is more nearly related to the arts of making masks and costumes.[15] To be sure, *opsis* is not optics, but it can hardly be denied that "staging" is fundamental to the effect—even the *affect*—of any tragedy, and Aristotle's apparent dismissal of acting and its supporting illusionistic devices tells us as much about the social status of mask makers and scene painters in antiquity as modern translations tell us about the presumed higher status of the verbal in relation to the visual in our own time. Neither of these ancient and modern presumptions should be taken to mean, however, that fundamentally important invention did not take place in the lower social strata of the highly esteemed and highly developed art of tragedy.

Skene most simply means "tent." (*Skias*, from *skia* [shadow], also means "shade, canopy, pavilion," so that the two are once again easy to confuse.) *Skene*, from which our "scene," "scenery," "scenario," and "obscene" (that which is properly offstage) all descend, referred to the temporary structure at the back of the stage permitting the entrance and exit of actors. According to Vitruvius, Agatharcus of Samos, who was self-taught, first painted the *skene* for a tragedy of Aeschylus.[16] Aeschylus died in 456 B.C., so, unless Agatharcus lived a very long life or there were two painters with this name, it might make most sense if the painting was done for a later presentation of his drama. In Aristotle's *Poetics*, we read that Sophocles (c. 496–406 B.C.E.) first used painted scenery.[17] A later source, Plutarch, tells us that when Agatharcus boasted to the younger and graver painter Zeuxis that he painted quickly, Zeuxis replied that his own painting both took a long time to make and lasted a long time once made.[18] This slighting rejoinder may be more important than it first seems because it suggests that Agatharcus's paintings were ephemeral, of little substance, or both. As we shall see, *skiagraphia* was associated with illusion and insubstantiality, but it could also be associated with painterly skills of technical abbreviation, with achieving illusion by the least means.[19] The anecdote may thus simply tell us that Agatharcus modeled the forms of his scene-painting, but it may also indicate that he was more highly skilled in the illusions of *skiagraphia*.

I will use one of the murals from the House of Augustus on the Palatine in Rome as an example of *skenographia* (figure 1.4). As we have just seen, Agatharcus probably painted his first *skene* toward the end of the fifth century B.C.E., and the murals in the House of Augustus were painted in Rome some four hundred years later, at the end of the first century B.C.E. I have two reasons for leaping this historical gap. First, there is a fairly consistent series of representations of *skenographia* from the fourth century. The fragment of an Apulian krater (figure 1.5) might be a compressed version of a tragic *skene*, showing a central episode with two flanking portals rather similar in delicacy and scale to the Augustan murals.[20] My second reason is that Agatharcus wrote a treatise on the art of painting about his invention, thus giving it another art-historical life.[21] And, still according to Vitruvius, the philosophers Democritus and Anaxagoras, following Agatharcus's lead, "wrote about the same thing, in order to show, if a certain central point is established at the apex of the extension of the rays of the eyes, corresponding to lines according to natural reasons, how the appear-

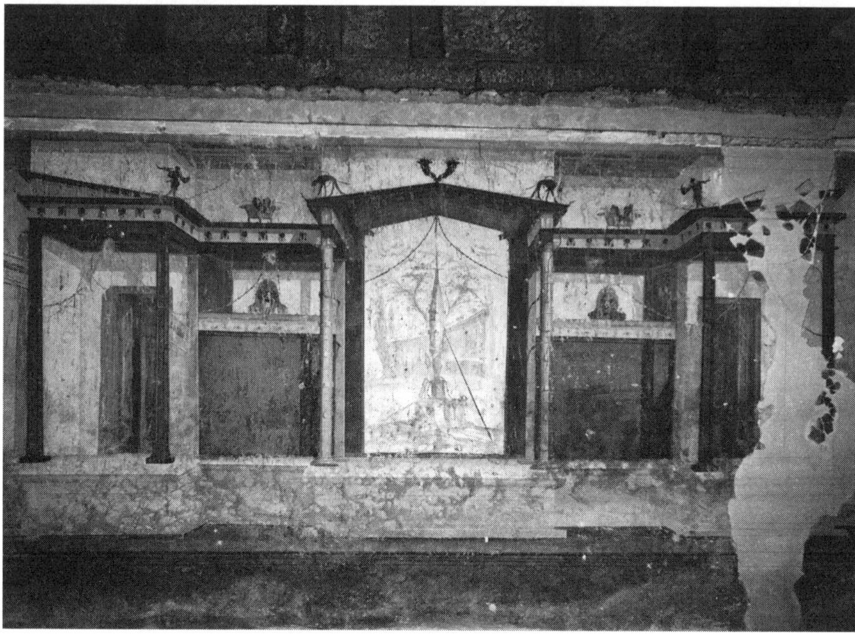

FIGURE 1.4. Tragic skenographia, *mural, House of Augustus, Room of the Masks, first century* B.C.E., *Capitoline, Rome. Photo: Deutsches Archaeologisches Institut, Rome (Koppermann, D-DAI-Rom 1966.0023).*

ance [*species*] may be made of certain images of uncertain things of buildings in paintings of scenes, and how that which is figured on an upright facing plane may be seen in some places to recede, in others to come forward."[22] If Vitruvius knew Agatharcus's treatise or its philosophical commentaries rather than just knowing *of* them, the painter of these murals in the house of Vitruvius's patron might also be supposed to have known them. The murals are precisely constructed with respect to a central point, which is highly unusual among the many Roman adaptations of Greek *skenographia* to interior decoration.

The *skenographia* in figure 1.4 displays the whole cluster of optical features, including highlights, the last of which, if Gombrich is right, would not have been available to Agatharcus himself. A painting of architecture like this could have begun with a front elevation, points on which were followed to one central point—here, the baetyl in the bright landscape, itself a high-tonality *skiagraphia*.[23] This construction yields a framework of recessive diagonal lines (and optical planes). The diagonal lines bound parallel transverse and vertical quantities in the virtual dimension, which, although equal, apparently diminish toward the line of sight because they are represented as if seen from a fixed

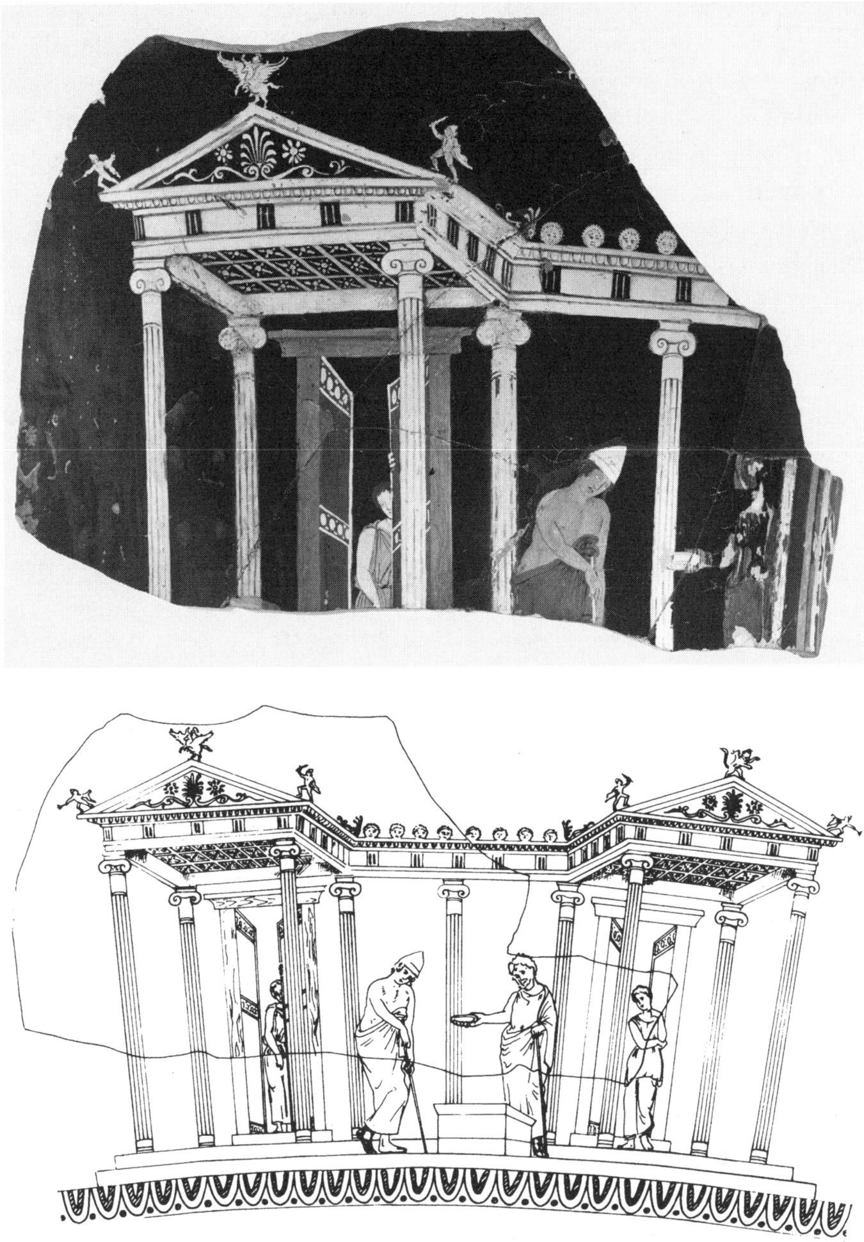

FIGURE 1.5. Tragic Scene, *fragment of an Apulian krater from Tarentum, ca. 360–350 B.C.E., Martin von Wagner Museum, Universität Würzburg. Photo: K. Oehrlein. Drawing: B. Otto.*

point along the same line of sight. The trapezoidal floor of the podium and the ceilings are examples of optical planes. A precise geometry of light falling from a source to our upper right intersects the delicate architectural members. Crisp shadows fall on the red panels flanking the central royal portal, and the thin yellow columns distinguishing this portal are modeled and highlighted in response to this light.

Modeling and shadows bring us back to Apollodorus of Athens—known as Skiagraphos, the "Shadow Painter." The meaning of *skiagraphia* is disputed,[24] and, as I have indicated, I will use the term primarily to refer to the modeling of virtual forms—that is, to the modulation of value (to use the modern term) within their contours. The modeling of a form must always imply a source of light, and his nickname might also have meant that Apollodorus was the first to show figures with cast shadows, which once again depend on the geometric economy of light. Perhaps because modeling implies shadows, we seldom notice when modeled forms in paintings do *not* have shadows, which most often they do not. There are good reasons for this. The "fall" of shadows may yield any number of anomalous shapes and obscure nearby forms, as I will discuss from time to time in the following chapters. Shadows may be cast on a wall, as they are in the so-called *Street Musicians (Scene from a Comedy)* (figure 1.6), but usually, as is also the case in this little mosaic, they fall foreshortened, and therefore distorted, on an optical plane—here, the stage on which the figures play and dance. As we have seen, shadows locate forms in virtual space, but they may also complicate or confuse pictorial composition. As Quintilian wrote, "paintings in which nothing is painted around figures [*circumlitus est*] do not stand forth," but "painters who put a number of things in a single painting make clear spaces between them, so that shadows do not fall from one figure onto another."[25]

Shadows are rare in the history of art because they may cause confusion but surely also because they have their own histories and mysteries.[26] In Hades, according to Homer, a ghost of the dead is usually called "psyche" or "eidolon" but may also be called "shadow" (*skia*), still highly mobile but forever incorporeal.[27] Our word "shadow" itself descends from comparable northern European terms but is also related to the Greek *skotos* (darkness, gloom, associated with death and blindness). When he defined "image" (*eikon*), Plato made no distinction between shadows and reflections, the latter of which he called *phantasmata* (to be seen on the surface of water and other dense surfaces).[28] If shadows are considered doubles or spirits, as they have often been, it might or might not be desirable to include them together with images, which themselves have related but generally more positive values. In a rationalized world—or, as we say, an

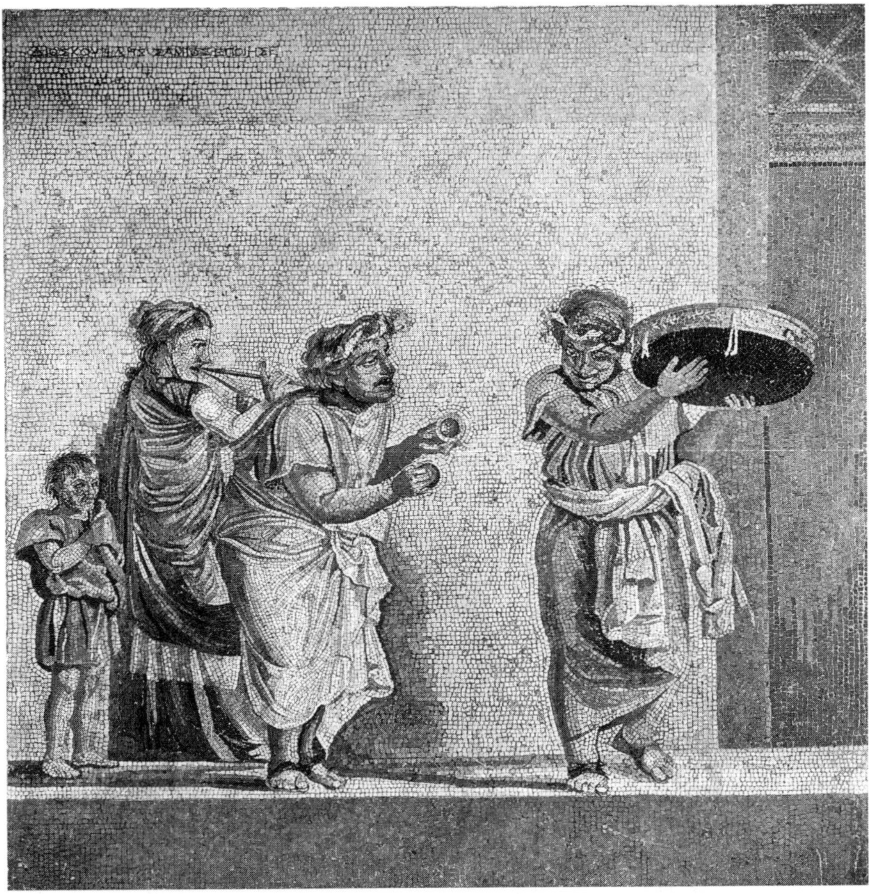

FIGURE 1.6. *Dioscurides of Samos,* Street Musicians (Scene from a Comedy), *mosaic from Cicero's villa, Pompeii, first century B.C.E., Museo Archeologico Nazionale, Naples. Erich Lessing/Art Resource, New York.*

"enlightened" world—shadows are regarded simply as consequences of the occlusion of light by opaque bodies, including our own. For the most part, however, the associations of shadows are, as we say, dark. We are inclined to think of the shadows of living things and especially of people as literally *of* them, as being in some genitive and even vital relation to them. Shadows thus have the anthropomorphic values of what might be called protean or polymorphous companions, mercurial, never detached while we live. Pliny's story of the invention of painting from the circumscription of shadow might thus have highly ambivalent and suggestive connotations and associations.

Pliny's brief life of Apollodorus is itself built around conceits of light.[29]

Earlier painters, Pliny said, were illustrious (*inlustres*), but their lives must be quickly bypassed in order to reach the lights of the art (*lumina artis*), among whom Apollodorus of Athens first "shone forth [*refulsit*]" in the ninety-third Olympiad (408–405 B.C.E.). He first expressed appearances (*exprimit species*), and it was he who first brought rightful glory to the brush. Pliny's choice of *exprimo, exprimere* (to press or squeeze out) should be noted. We are accustomed to thinking of art as expression, but it would of course be anachronistic to suppose Pliny intended any such modern meaning. The second definition of *exprimere* is "to press upward or elevate, to form by pressure, to model or shape." Most generally, *exprimere* means "to imitate," and here, it may be supposed, specifically "to imitate appearances." Putting all of this together, we might understand Pliny to say that Apollodorus, by modeling forms, had imitated *species, phantasmata*, things as they are seen, even perhaps things as they seem, so that "Shadow Painter" might mean painter of the fleeting and insubstantial in a new and positive sense. Apollodorus had begun a way of imitating the world of things in light as they come to be and pass away before our eyes, gone before we know it. At the same time, Pliny concluded with the words I emphasized in the introduction: "there are no paintings to be seen prior to those of Apollodorus that hold the eyes [*teneat oculos*]."

In an earlier section on colors, Pliny had provided a broader framework for the fundamental importance of Apollodorus's breakthrough. The first paintings, we have seen, were monochromatic. Eventually, Pliny said, "art differentiated itself [from the monochrome] and invented light and shadow, the alternate differences of color making them stand up together. Later, highlight [*splendor*] was added, which is something quite different from light [and shadow]. The specific relations between light and shadow they call *tonon* and the commeasured transitions of colors they called *harmogen*."[30] Without naming Apollodorus here, Pliny has reviewed the same crucial transformation in the art of painting he attributed to Apollodorus.

The new painting begun by Apollodorus thus used oppositions of light and dark within the contours of forms, and the words "*tonos*" and "*harmogen*" suggest that right relations between light and dark were to be found by a process characterized as one of tuning and fitting. All of this preceded highlights (*splendor*), which as we shall presently see, are essentially reflective. In any case, the possibilities opened up by his innovation were vast. Apollodorus himself painted a priest at prayer, which might be supposed to have been serene enough, but he also tackled a more ambitious theme, *Ajax Struck by Lightning*, in which light effects must be supposed to have been—as we still say—"sensational."

The masterpieces of Greek painting—including those of Apollodorus—are long since lost, and the works of painters once world renowned, works commanding royal and imperial prices, are simply no longer to be seen or are known to us only from descriptions or in late, pale reflections. Writings on art, including some on color, are also almost entirely lost, together with the actual results of the practical resolutions of these theories and oral traditions of instruction in what were certainly most highly developed technical traditions. Even within the severe limits set by the nearly complete destruction of ancient painting, however, I believe it is possible to outline its advances in very general but still useful terms.[31]

The dancer from the Etruscan Tomb of the Triclinium at Tarquinia (figure 1.7) may be taken as an example of the early painting Pliny called "monochrome." The figure is not modeled, and with adjustments for the formats involved, the treatment is very familiar from Greek black-and-red-figure vase painting, which is often taken to stand for early Greek painting taken altogether. The next stage, representing the innovations of Apollodorus at some remove, is exemplified by paintings in the *tholos* at Kazanlak from the end of the fourth century. In the detail of a groom in figure 1.8, it first appears that there is no more than a thickening of contours, but if we follow, for example, the contour of the higher shoulder to the neck, it is clear that shadow is intended, and so throughout the figure. The ribbons of dark in the drapery are something like the deep shadows made by the running drill in the revolutionary Phidian sculptures of the Parthenon (figure 1.9). This groom's clavicle, which in earlier Greek painting would have been shown by a locating and identifying line, is now a series of short parallel marks defining it as a lighted (or shadowed) surface. The prominent hatchings—the word itself descends from words meaning "to chop or cut" and so preserves an analogy to sculpture—are like the chiseled transitions from light to drill-cut shadows in Phidian sculpture. They are literally halftones, and throughout the figure, hatching effects passages from shadow to color. There are no added highlights, and the variations I have described were accomplished with dilutions of earth brown, an earth red, and the white of the plaster wall.

The formula evident in figure 1.8, rather than simply being replaced by the changes I am about to describe, had a long historical life in its own right. The general principle—Gombrich's "rule of Philoponus"—is that light is nearer, dark farther away; it is as if three-dimensional forms advance toward us from shadow to light or, to look at the matter a little differently, as if a viewer's eye were the source of light and its action weakened over a short distance, which approximates Plato's extromissive theory of vision.

There was some disagreement in antiquity over who deserved credit for the

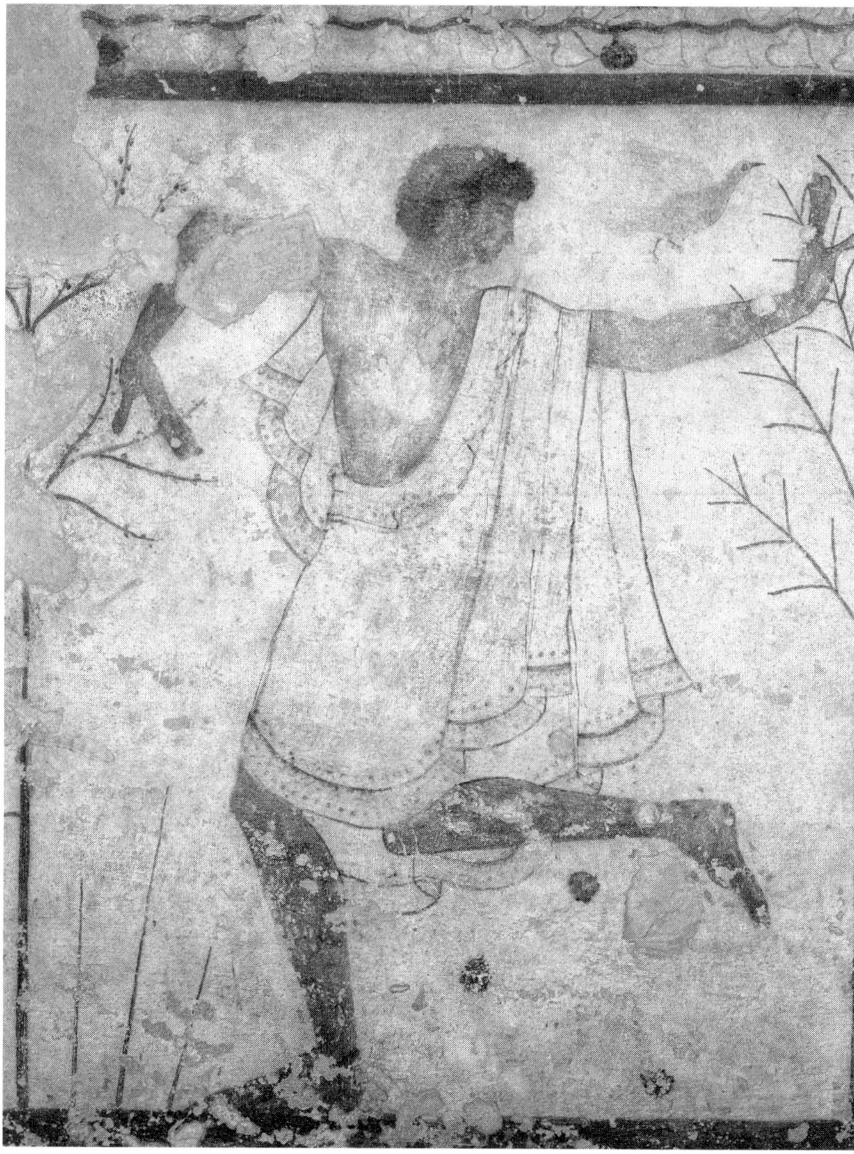

FIGURE 1.7. Male and Female Dancers, *Etruscan, Tomb of the Triclinium, Tarquinia, detail, ca. 480–470* B.C.E. *Scala/Art Resource, New York.*

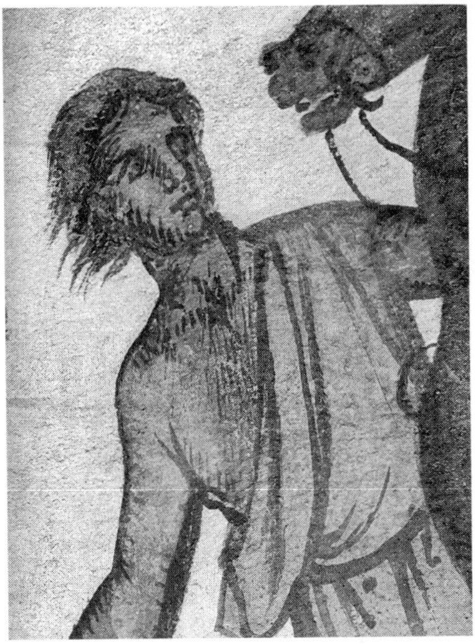

FIGURE 1.8. Groom Leading an Empty Quadriga, *detail of mural, tomb at Kazanlak, Bulgaria, ca. late fourth–early third century* B.C.E. *From Ljudmila Shivkova,* Das Grabmal von Kazanlak: Farbaufnahmen Julian Tomanov *(Recklinghausen: Bongers, 1973), fig. 11.*

momentous change in the history of painting signaled by the appearance of modeling. Plutarch agreed with Pliny, calling Apollodorus "the first man to discover the art of mixing colors and modeling of shadows,"[32] but Quintilian said that Zeuxis "invented the method of light and shadow [*luminum umbrarumque invenisse rationem*]."[33] Apollodorus himself, wispy as his historical person might be, seems not to have been silent on this question. Once again according to Pliny, Apollodorus wrote an epigram saying that Zeuxis (to whom Agatharcus had boasted of his speed of execution) had stolen the art of painting from its inventors and carried it off. This could simply mean that Zeuxis had imitated his innovation, but it might also mean that he picked up where Apollodorus left off, then brought painting to new heights. However that may be, Pliny's next paragraph begins grandly: "The gates of art having been thrown open by Apollodorus, they were entered by Zeuxis of Heraclea."[34]

We may perhaps get some sense of this accomplishment in the figure of Rhadamanthus, ruler and judge of Elysium, from the facade of the chamber tomb at Lefkadia from the turn of the fourth and third centuries B.C.E. (figure 1.10). Here the scheme of Apollodorus is magnified and refined in much more coloristically varied and composed surfaces. More importantly, the source of light is treated in a fundamentally different way. To our right is once again the thick shadow contour, but no such contour is to be seen to our left, as if the source of light were not in our eyes but rather somewhere to our left.[35] If this painting is to be associated with the style of Zeuxis, it was he who painted that

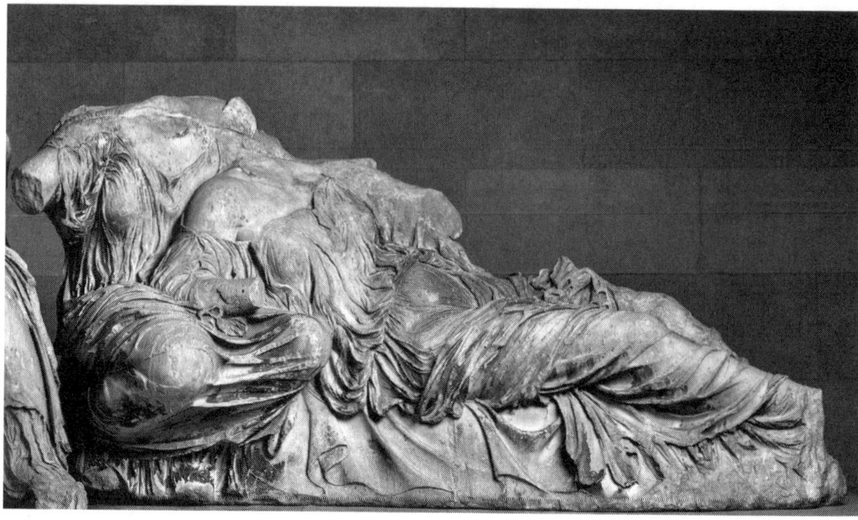

FIGURE 1.9. Aphrodite Reclining in the Lap of Her Mother, Dione, *Parthenon, east pediment, ca. 438–431 B.C.E., British Museum, London. © Trustees of the British Museum.*

paradigm of Western illusionism, the grapes that fooled the birds; who gathered the five loveliest maidens of Croton in order to paint his Venus or Helen of Troy;[36] and who, according to Quintilian, "amplified the members of the body, deeming that [amplification] fuller and more august," following "Homer, who liked to make all figures, including women, more robust."[37] In this expansion of scale, Zeuxis might also have followed Phidias, who might himself have followed such a reading of Homer. Zeuxis might also have learned from Phidias the complexity of the surfaces of his forms in light (figure 1.9).

It is at this point in the bare history of Greek painting we have been left that a writer appears whose characterizations of painting have long outlasted any examples of the art. Both Agatharcus and Apollodorus were contemporaries of Plato (427–347 B.C.E.), who at the beginning of the fourth century B.C.E. would have been in his late twenties. (Socrates drank the hemlock in 399.) If, as has been argued, Plato is the first writer to use the word "mimesis" to refer to the imitation of appearances, perhaps this in itself echoes conversations and controversies with regard to the new painting.[38] Since Greek painting was always figural, it would have been difficult to separate depictions of events (as in earlier vase painting) from their representations as if seen—that is, as optical. *Skiagraphia* would then have been a special case of mimesis in the older sense of enactment or mimicry. It would, however, have been a crucially important break, making painting mirrorlike, as Plato himself clearly recognized. In the *Theae-*

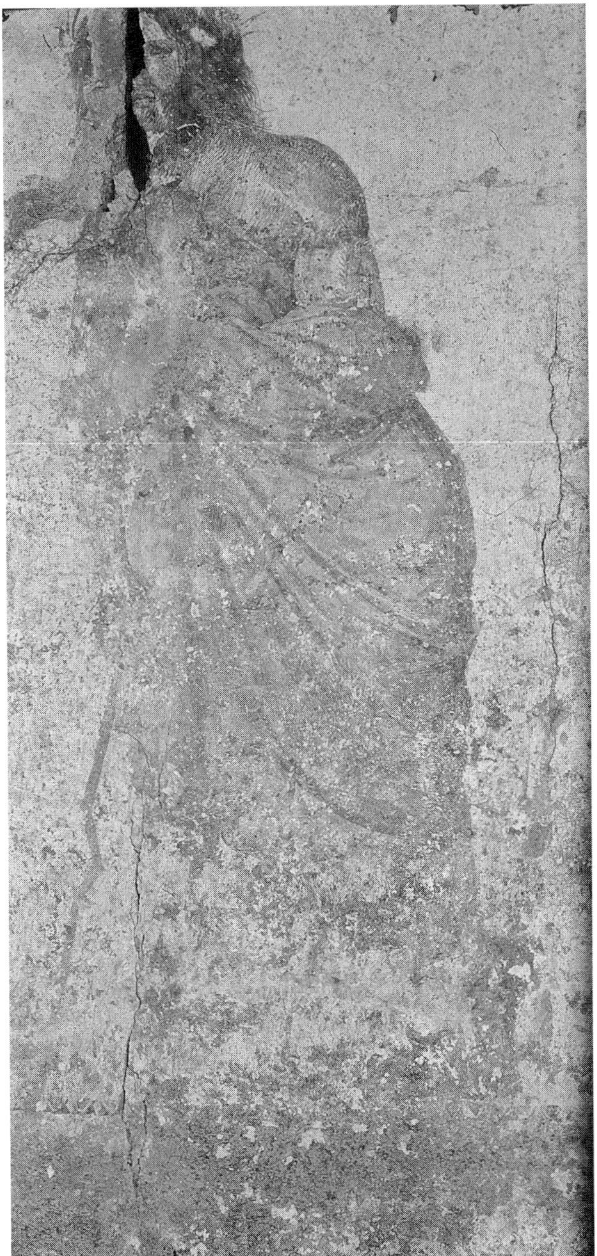

FIGURE 1.10. Rhadamanthus, *from the Great Tomb at Lefkadia, late fourth century* B.C.E. *© Archaeological Receipts Fund, Hellenic Republic, Ministry of Culture. (Reproduced in color as plate 2.)*

tetus, Plato says that an argument is like a *skiagraphia*, a shadowed form that seems to be something only at a distance.[39] In the *Republic*, he added a slippery temporality to this illusion of light and dark, likening painter and sophist to someone carrying around a mirror, instantly making images of sun, sky, earth, himself, and everything else but only the appearance (*phainomena*) of all these things, not the truth.[40]

Plato's arguments against the illusions of sense, sophistical rhetoric, painting, poetry, and theater are basic to his philosophy; they also have the urgency of polemic about them. The run of humanity, "grazing and copulating," as Plato put it, pursue what in his famous allegory of the cave in the *Republic* he again and again calls "shadows," yielding to "mad desires." *Skiagraphia*, rather than curing our soul, exploits its weakness, showing us the surfaces of everything, the substance of nothing, shadows in a most negative sense.[41] In the *Sophist*, Plato compared painting to rhetoric. What is to be made of arts claiming the power to imitate everything? In reality, both painting and rhetoric, by showing pictures at a distance, will be able "to deceive the duller ones among young children."[42] *Skiagraphia* plays on our passivity to phenomena, just as do all kinds of clever contrivances (*mechanai*) and sleights of hand, appearances the causes of which we do not understand, the merely marvelous aspect of skill.

In the *Protagoras*, Plato, at least as emphatically as in the *Republic*, makes a distinction between what he calls the "art of measurement" and the "force of phenomena [*phainomenou dynamis*]," the force of what appears in the light.[43] The first leads us toward truth, the second into error, and again, it is the second that illusionistic art imitates. Plato's repeated example is one thing seen as large and small, like identical columns receding in a colonnade. For him, this apparent difference in identical quantities implies contradiction and paradox in sensation itself, and its best consequence lies in urging the higher faculties of the soul to verification by measure and to such properly dialectical issues as identity, true difference, and the nature of dimensions as such. Plato insisted that those who think that pleasure is relative to pain are like those who, ignorant of white, contrast gray with black:[44] they cannot know true pleasure, which is that of the mind. *Skiagraphia*, which Plato included with diminution and refraction, is "colored by contrary juxtapositions as to seem intense in either kind" and makes us see concavities and convexities when there are none, playing on our proneness to error in the judgment of surfaces.[45]

Plato was not alone in his distrust of shadow-painting, and the Augustan historian and critic Dionysius of Halicarnassus used the example of *skiagraphia* to set "before the eyes" the difference between two orators. "There are some old paintings which are worked in simple colors without any subtle blending of tints but clear in their outlines, and thereby possessing great charm; whereas the later paintings are less well-drawn but contain greater detail and a subtle interplay of shadow and light and are effective because of the many nuances of color which they contain."[46] Dionysius made the same comparison between Lysias and Isaeus, the latter of whom is more artful but also fails to conceal his art. In this case, *skiagraphia* and the variation of color it makes possible parallel the fantasy rejected by Horace and Vitruvius,[47] displaying both classical taste and a "preference for the primitive."[48]

For Plato, just as the ears were framed for harmony, for what they most desire to hear, so the eyes were framed for astronomy, for surveying the soul-healing order and regularity of the heavens.[49] When he imagined the education of his guardians, Plato began with arithmetic and geometry, moving on to astronomy. But then he backtracked, saying that before studying astronomy it was necessary to learn the still undeveloped field of solid geometry, of which optics, with its visual "cones" and "pyramids," might be considered a subdiscipline. In the late *Timaeus*, Plato identified the highest element of fire with the pyramid and described sight as the action of a subtler fire emitted from the eyes, capable of joining with the natural fire of light. This, however, is far from a developed optics. If Plato's writings more or less coincide with the appearance of illusionistic painting, then according to the thesis of this chapter, they also coincide with the beginnings of geometric optics, of which Plato might be supposed to have been deeply suspicious. His pupil, Aristotle, much more favorably inclined to physical geometries, was said to have made great advances in optics (of which there is now little trace), and contributions to the emerging science must have been ongoing before its formulation by Euclid around the end of the fourth century.

Rather than equating illusion with vision and therefore vision with illusion, as Plato had done, Aristotle made a distinction between things that are false by nature, like shadow-paintings and dreams, and things that cannot be substantiated. There really *are* shadow-paintings and dreams, but their reality is not that of which they create the appearances.[50] These appearances have their own reasons, and this raises the prospect of the imitation not just of *what* we see, but also of *how* we see. The idea of optics as a physical geometry thus has the most important implication that there are reasons to be found in the paradoxes of vision. We see the successive columns in a colonnade to be smaller and smaller because they are seen under ever-smaller visual angles. We may thus, in the words of Vitruvius, find certainty in uncertain things,[51] and the contradictions of vision to which Plato continuously pointed may be explained. Of course, when the reasons of vision are understood, it means just as surely that

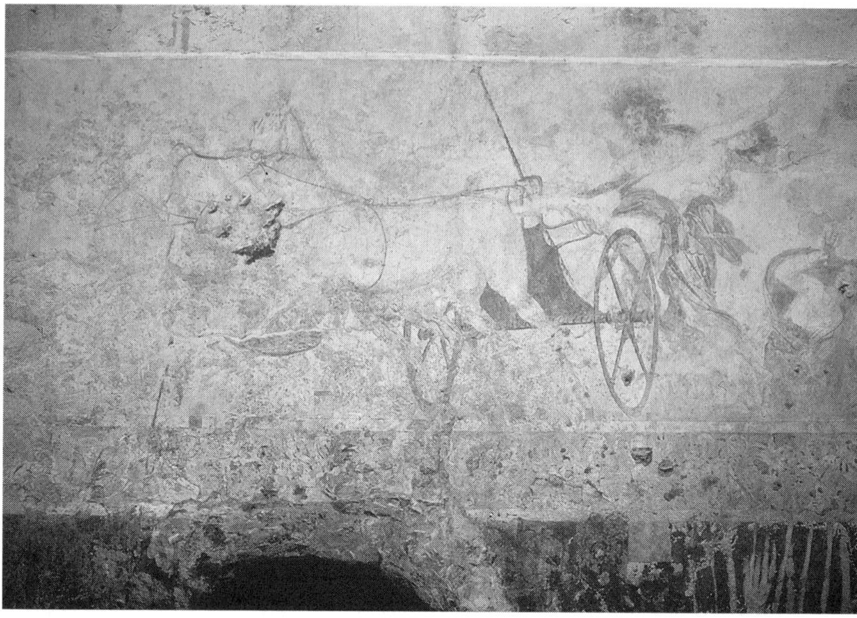

FIGURE 1.11. Abduction of Persephone, *from Tomb 1, Vergina. © Archaeological Receipts Fund, Hellenic Republic, Ministry of Culture. (Reproduced in color as plate 3.)*

appearances and illusions may be constructed according to them. Furthermore, if actual appearances may be imitated by construction—if a long series of columns may be made to seem to recede on a two-dimensional surface—then the utterly invented and fabricated may also be given something like the force of phenomena and the fantastic may be presented as if seen.

With this in mind, we may return to the history of Greek painting. The superbly drawn *Abduction of Persephone* from the great tumulus at Vergina (figure 1.11), attributed to the quick-painting Nicomachus or to his pupil, Philoxenus of Eretria, who devised ways of painting even more rapidly, displays complete command of the earlier scheme of modeling and light from a source. The figures are almost entirely made up of fluidly foreshortened forms with modulated surfaces lighted from our left; contours on this lighted side are broken, as if themselves surfaces shifting and shimmering in light. This painting also brings us to the Macedonian court, once again to the vicinity of Apelles, and to the completion of the system of optical naturalism I have been describing.

The highlight, commonplace as it may seem, takes us from optics to catoptrics, the geometry of reflections.[52] Highlights in fact *are* reflections, and in the ideal case of a perfectly smooth surface, they are governed by another simple principle: the angle of incidence equals the angle of reflection. This principle derives once again from the economy of light, just as the visual angle of optics does, and is like the visual angle in that it necessarily locates a viewer both in relation to a surface and to a source of light. When we move, reflections on a surface move—assuming a stationary source of light—always according to the same rule. (The same rule of course applies if the source of light is moved, and effects vary according to the texture of the object.) Highlights once again—and even more emphatically—define vision as a physical relation, the relation of a viewer to the actual world, and since surfaces, which are always one or another texture, reflect light in characteristically different ways, highlights may be said to specify in a new way the fusion of sight and touch implicit in the modeling of virtual forms.

According to Pliny, Apelles was the student of Pamphilus, who, Pliny also tells us, first practiced painting as a liberal art. Pamphilus was "the first in painting to be learned in all letters, especially arithmetic and geometry, without which he denied the art could ever be perfect."[53] Pamphilus accepted no one for instruction for less than a talent, and he made drawing part of aristocratic education in "all of Greece."[54] Apelles paid Pamphilus's steep price for instruction and in turn wrote volumes on painting, now lost. There is no reason to suppose that the ideal of the learned artist did not precede Vitruvius's learned architect or that "arithmetic and geometry" would not have included optics.

The large mosaic of the *Battle of Alexander the Great and Darius* from the House of the Faun at Pompeii (figure 1.12) (about 17.5 by 10.5 feet including the frame) is a meticulous copy after a late-fourth-century painting sometimes attributed to Philoxenus of Eretria (perhaps the painter of the *Abduction of Persephone* [figure 1.11]), sometimes to Helen of Egypt, and most recently to Apelles.[55] The mosaic must have been made from a copy of the painting, and, given this double transmission, it is impossible to know at what point misapprehensions or approximations might have crept into some of its complex optical and spatial effects. Most of these tour de force passages are treated with assurance, but it might be supposed that some details—the treatment of the wheel of Darius's chariot, for example—were not understood. The battle, framed by illusionistic dentils apparently lighted from the front center, again as if by the point of sight of a viewer, takes place on a stage, a segment of a sandy plain with a few stones and stumps and a dead plane tree in the near background. The sky is white, presumably with the dust of battle.

Apelles was the greatest of the painters who, according to Pliny, worked in

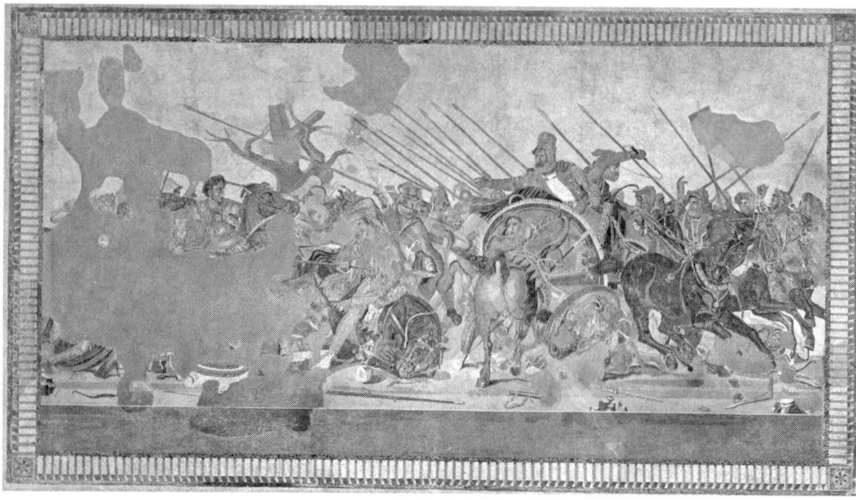

FIGURE 1.12. Battle of Alexander the Great and Darius, *Roman mosaic from the House of the Faun, Pompeii, second–first century* B.C.E., *after a fourth-century original by Apelles(?), Museo Archeologico Nazionale, Naples. Alinari/Art Resource, New York. (Reproduced in color as plate 4.)*

the "noble" manner using only four colors— white, yellow ochre, sinopia red, and black (which, it may be noted, are listed from bright to dark between white and black).[56] As for the depicted light in the mosaic, it falls from our upper left, from above the head of Alexander, and may perhaps best be judged in the broken lance and its shadow to our lower right. The precision and consistency of the description of the fall of light allowed Paolo Moreno to conjecture as to the time and direction of Alexander's charge. Such cosmic location of a crucial event may have been appropriate to Alexander, but it also implies that Apelles not only understood optics and catoptrics but also had a working understanding of closely related geography and astronomy (and perhaps astrology).[57]

Moreno's argument implies an embracing development of physical geometries that provides ancient precedent for the specialist culture for which I will argue in the following chapters on the Italian Renaissance. It will be recalled that Vitruvius devoted the ninth of his ten books on architecture to astronomy, with an eye to the making of sundials and clocks, and the same cosmological geometry might have been used to specify the place, time, and light of Alexander's victory. A late classical source, the *Saturnalia* of Macrobius, recounted the atomist theory of vision of Democritus and Epicurus, which was rejected, however, in favor of the theory of Plato.[58] Macrobius also described the visual angle, spreading from the point of the eye to the height of the heavens, as the usage of painters, and he defined the meridian, the zenith point of the sun, which varies from one place to another. If Apelles meant to show the time of the battle, he must have known the meridian of the place it occurred as well as the time of year. Macrobius also defined the limit of vision, what "the ancients called" the horizon, the circle of which we are the center wherever we go. This circle cannot exceed a diameter of about thirty-five miles, at which point the line of sight becomes tangent to the earth's curvature and sight fails. Such arguments might echo attempts to define the size of Earth's sphere by measuring its curvature relative to an ideal case of sight. However that may be, they expand physical geometry to cosmic dimensions and scale, which would not only have commemorated Alexander's victory but nested it in the orderly turning of the universe itself.

Pliny tells us that another student of Pamphilus, Pausias, first invented foreshortening, which Pliny rather awkwardly described as follows:

> first, when [Pausias] wished to show the length of an ox [he had painted a large *Sacrifice of Oxen*] he painted it facing the viewer rather than in profile, and [yet] its great size is understood; then [second], whereas all [painters], when they wish eminences to be seen make brighter colors, and make those things black they wish to recede from sight [the "rule of Philoponus"], [Pausias] made the whole ox black, and [thus] from shadow itself gave body, with great art showing [things] to stand forth on a plane [surface] and to be entirely solid, even though partial.[59]

Pausias's ox was "partial" because some foreshortened forms must have occluded others. A foreshortened form is successful when its drawing makes us imaginatively complete occluded parts; since the ox was all black, its three-dimensionality could not have been shown by modeling. Light-dark contrast—and relative eminence, standing forth—could, however, be achieved by another means, by highlights, by treating the black surfaces as glossy and reflective. Pausias had thus achieved something marvelous and paradoxical, showing three dimensions in two, making body of shadow—certainly the two are meant to be opposites, like something of nothing—and making visible parts suggest invisible parts, and suggest the whole.

If Apelles did not invent foreshortening, he quickly mastered it. For the Temple of Artemis at Ephesus, one of the wonders of the ancient world, he painted Alexander holding a thunderbolt, in which "the fingers seem to stand forth, and the thunderbolt to protrude from the panel." Apelles also made paintings of horses, one of which was said always to have made real horses neigh in recognition.[60] Among the foreshortened horses in the *Battle of Alex-*

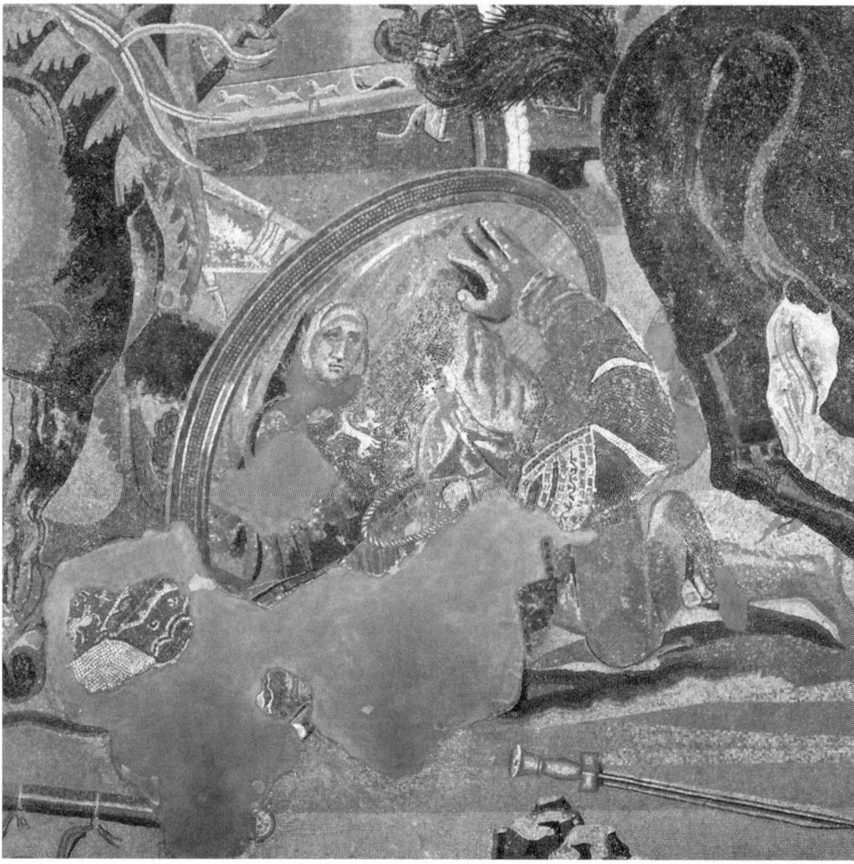

FIGURE 1.13. Battle of Alexander the Great and Darius, *detail of figure 1.12, a fallen warrior and his reflection in a shield. (Reproduced in color as plate 5.)*

ander the Great and Darius, the horse to our right, the nearest of the four frightened horses pulling the chariot of Darius, is black. The surfaces of this horse are not simply modeled, however; they are instead reflective, and we see them with a kind of muted sheen, light reflecting from the animal's sleek hide, perhaps glistening with sweat. The horse most foreshortened of all, shown rump-first beneath the chariot of the Persian king, his neck coiling in terror away from the onrushing Greeks, is brown and modeled more nearly according to the usual formula, turning from light through brown to black.

As I have said, all highlights are reflections, and in this scene there are many more reflective surfaces than simply modeled surfaces. Certainly the most spectacular reflection is to be seen in the polished shield beneath the wheel of Darius's chariot (figure 1.13). This great round disk, perhaps meant to be of polished silver, is also foreshortened. Its bearer—we see his legs to our right of the shield, a bit of his curly hair to our left—has fallen, and in this instant the shield topples over a second fallen figure in front of it, his left leg extending to our left, under the hooves of the galloping horse. White highlights reflect the light of the sun from the mirror's rim to our eyes, and the surface of the shield is shown more or less precisely as a reflection in a convex mirror. We might presume that the painter simply imagined this reflection or had seen such reflections, but it is much more likely that he (or she—it still might be Helen of Egypt) understood how reflections work and made that understanding part of the practice of the art of painting. In the mirror of the shield, against a still-dizzying reflection of the earth onto which the figure falls, the dark underside of his hand is reflected at the size of the hand itself. It is not clear that the copyist or mosaicist entirely grasped the transition from hand to face, but, as should be the case in a convex mirror, the reflected face is relatively smaller, and it is this smaller visage that turns its sad Skopasian eyes to the shadowed hand.

This chapter has been made up of reconstructions in imagination that can only dimly recall for us the utterly lost works of the great painters for the great patrons of antiquity. These reconstructions may also remind us that the ancient "imitation of appearances" must have been rich and complex in unique ways and that our imagination of ancient painting must suffer from the simplicity of our own, inevitably photographic, understandings of this phrase. As the tradition of ancient painting continued to unfold, optical themes continued to be repeated and varied, but they also continued to provide the fertile basis for invention and observation, for the emergence of new problems and their solutions in new skills. Full mastery of the optical devices I have discussed, together with reflection into shadows, is apparent in the radiant little mosaic with the signature of Dioscurides of Samos from the first century B.C.E. (figure 1.6). And of course the "painterly" abbreviation of figures and landscape, finally based on the same abstraction of visible surfaces into contrasts and modulations of light and dark, proceeded apace, becoming the routine craft but also the always poignant phantoms of Roman painting.

I wish finally to return to the question with which I began. Why were the light-modeled paintings of Apollodorus of Athens the first to "hold the eye"? One ancient answer might have been that they did so because they were sightlike, not like appearances so much as like sight itself. But to return to my original point, making paintings sightlike presupposed an extraordinary abstraction of vision. This analysis—or untying or unknotting—of the bright manifold of vision solved paradoxes but at the same time made any number of new brilliant and fantastic syntheses possible and credible. Plato did not simply disapprove of the imitation of appearances, weak and at a second remove from their perfect and shadowless formal principles; he also worried about the fascination and dangers of the irresistibly marvelous and psychagogic force of art, which could give something like the force of phenomena to the fictions, dreams, and visions of the poets if also to the falsehoods of the sophists.

CHAPTER 2 *Speculum Animatum*

The optical features I discussed in the last chapter persisted in different ways and were adapted to the various purposes of western and eastern Mediterranean art through the thousand years between the close of Roman antiquity and the beginning of the Renaissance. While these features persisted, their references were no longer descriptive, nor did they function as schemata for the continued interrogation of visual experience itself. A Spanish Romanesque *Adoration of the Magi* (figure 2.1) displays modeling, highlights, and a version of the optical plane, as does an early-thirteenth-century Byzantine *Madonna and Child* (figure 2.2), but they do so in different ways. Neither image is thinkable without the distant precedent of ancient painting, but the pictorial transformations of ancient optical features they and countless other images exemplify took place through the centuries quite apart from the science of optics, which dwindled away in the early Western Middle Ages. In the sixth and seventh centuries, for example, Isidore of Seville, whose encyclopedic *Etymologies* was much consulted in the millennium to follow, briefly described extramission and wrote that the power of sight resides in the central pupil of the eye, where small images are to be seen.[1] When Western European painting again drew closer to optics in the late Middle Ages, their new union was determined at the deepest levels by changes in the purposes for which paintings were made and by advances in the science of optics, which had begun to assume its modern form.

The geometric optics paralleling the rise of Greek painting I described in the last chapter is very basic, and its principles did not go beyond those to be found in Euclid. But optics was much developed and expanded by writers after Euclid, first by Ptolemy in the second century, then in Islamic philosophy and science. The late-tenth- and early-eleventh-century philosopher Abu Ali al-Husain ibn Abdallah ibn-Sina, vastly influential in the West as Avicenna, argued for Aristotle's intromissive theory of vision and repeated the metaphor giving this chapter its title, comparing the eye to a mirror.[2] The Islamic study of vision culminated in the great optics of Avicenna's slightly older contemporary, Ibu Ali al-Hasan ibn al-Hasan ibn al-Haytham, called Alhazen in the West. Alhazen's *Book of Optics* was translated into Latin with the title *De aspectibus* in the late twelfth or early thirteenth century and by the fourteenth had been worked into scholastic variants and adaptations of older Neoplatonic and Aristotelian phi-

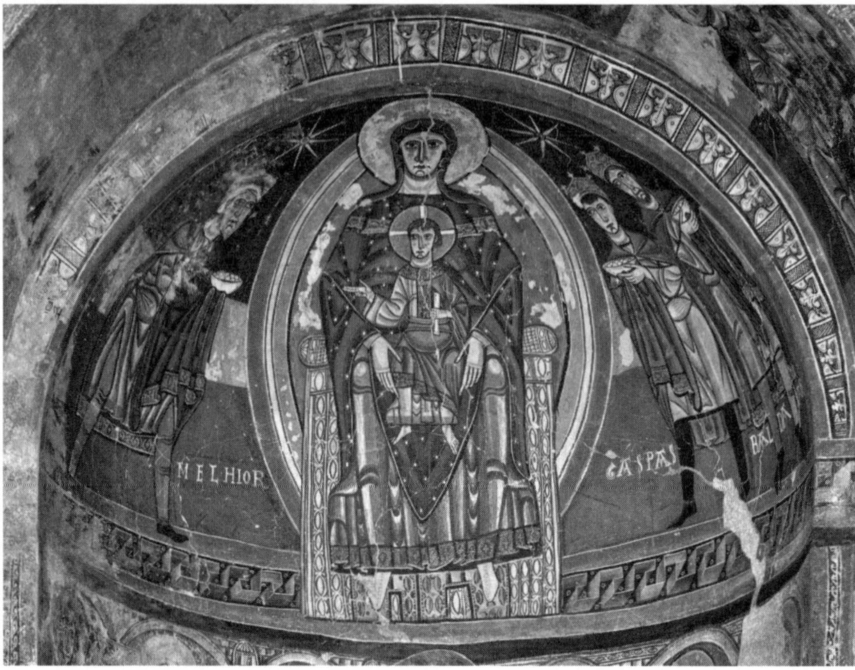

FIGURE 2.1. Adoration of the Magi, *from Santa Maria de Tahull, twelfth century, Museu Nacional d'Art de Catalunya, Barcelona. © MNAC–Museu Nacional d'Art de Catalunya, Barcelona.*

losophy, theology, and psychology.[3] Latin optics, the most influential of them closely based on Alhazen, appeared in rapid succession. By the later thirteenth century, when the gap between painting and optics had begun to close, the "art" of optics was both highly developed and widely accessible. The imagery of Dante's *Divina commedia* is inspired at many points by the new optics and its correlative cosmology, and this great poetry and its commentaries provided a familiar and fertile source for artists of the Florentine Renaissance.[4] The question to be addressed, then, is why it came to be thought necessary or desirable to close the gap between optics and painting, and as the investigations, demonstrations, and inventions made possible by this reunion proceeded, why it became possible for European artists to participate in what I called in my introduction a common specialist culture, playing across an enriched field of optics, catoptrics, cartography, and mechanical arts, in response to contemporary demands of many kinds.

In this chapter I will be mostly concerned with the historical appearance of painter's perspective, the invention of which has long been attributed to the

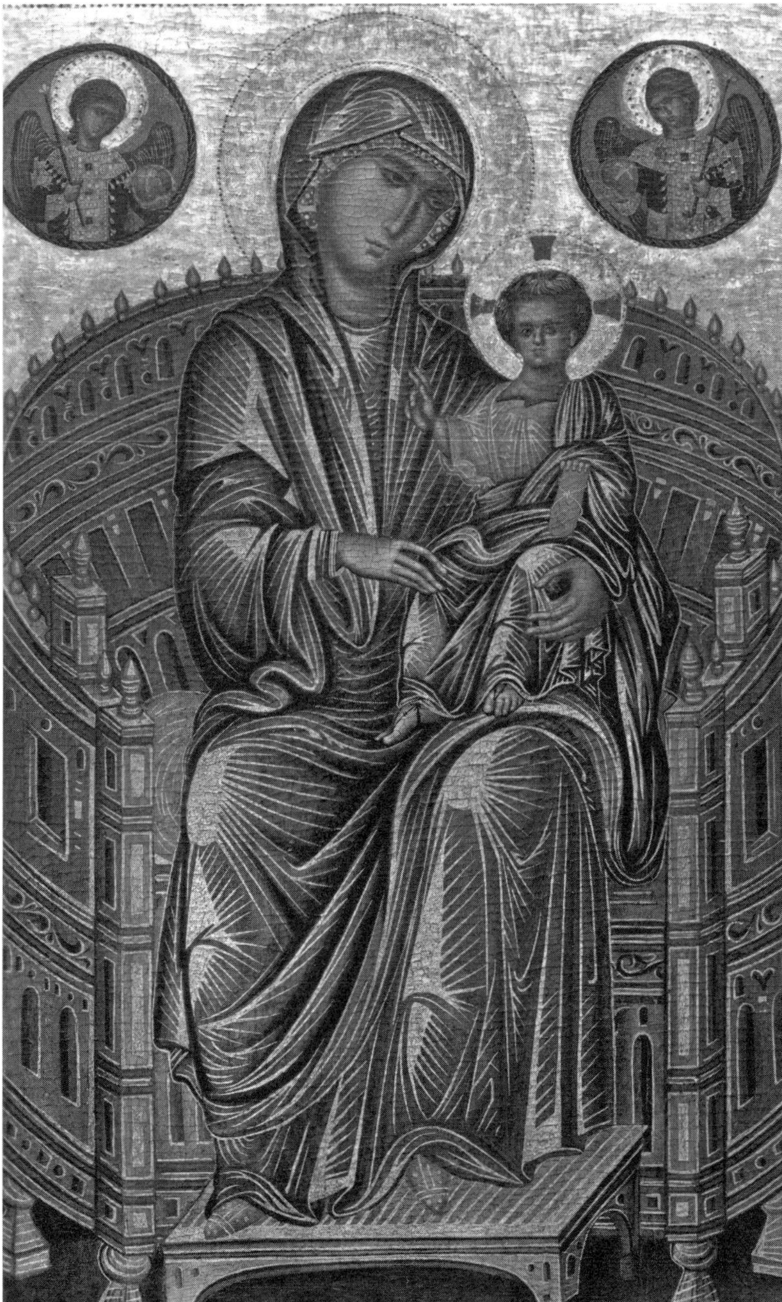

FIGURE 2.2. Madonna and Child on a Curved Throne, *tempera on panel, Byzantine, early thirteenth century, National Gallery of Art, Washington, D.C. Andrew W. Mellon Collection, © 2006 Board of Trustees, National Gallery of Art, Washington, D.C.*

Florentine goldsmith and architect Filippo Brunelleschi. No one denies that the advent of painter's perspective was crucial in the history of European art, and Brunelleschi's innovation has continued to be the object of much more scholarly and critical scrutiny than I will be able to consider. Modern attitudes toward perspective are mixed. On the one hand, there is a positive (if unreflective) tendency to fold perspective into a broader narrative of progress, treating it not as an "invention" at all but rather as a discovery. According to this view, the laws of vision were discovered like the law of gravity, and perspective ended up on the side of science, of the objective as opposed to the subjective. This general view made it possible to dismiss traditions of art that had not "discovered" perspective but at the same time to oppose perspective to the subjective roots of art itself. Perspective was consequently a major casualty of the rise of Western artistic modernism, assuming more negative connotations of rationality and mechanism, placing it altogether outside the bounds of "art." In more scholarly discourse, perspective has become a cultural symptom, a "symbolic form" in Erwin Panofsky's adaptation of Ernst Cassirer's neo-Kantian phrase,[5] and it is at this level of generality that perspective has been argued to be an arbitrary Western representational code,[6] or an early and ominous manifestation of modern instrumental reason and panopticism.[7] I will argue outside these alternatives that painter's perspective is a special case of the long tradition of geometric and physical optics, related to scrying on the one hand but also to telescopy and microscopy on the other—integral, that is, to magic and mystery as well as to the rise of natural science, culminating in the mechanization of vision in photography and in the animation of mechanized vision in film and television. Perspective is thus related to modern metaopticality[8] and so to methods of control and surveillance, but it has also given us some of the characteristic metaphors of modernity (perspective, point of view, even relativity) and thus represents a substantial contribution to the history of the formation of the modern Western subject—and *idea* of the subject—an issue to which I will return in my afterword.

Paul Hills noted the coincidence between the presence at the papal court in Viterbo during the 1260s of three major writers on the new optics—Roger Bacon, Witelo, and John Pecham—and the revival in Rome of monumental painting (and mosaic) under the patronage of the Orsini pope Nicholas III.[9] In this painting, a simplified and emphatic treatment of the modeling of virtual forms is joined with a newly theatrical space and dramatic treatment of theme. To be sure, this revival looked to precedents in late classical and early Christian art, but the new optics was still based on its ancient principles, one of which was, as Alhazen phrased it, that "light and color are perceived in themselves by sight."[10] If we see at all, we see color in light, and if this definition of the visual is put together with the Aristotelian (and Scholastic) principle that there can be no knowledge without the mental images given in sensation,[11] then the science of vision assumes new dimensions, even a new absoluteness. The elements of vision are shared by all of us, and sight thus makes a universal appeal, an ecumenical appeal. The pronounced and methodical modeling of the new painting to be seen in the painting of Pietro Cavallini and Giotto is consistent with this fundamental optical principle and, more importantly, is consistent in new ways with the ancient doctrine of the Church that images should be exemplary and instructive. In its Aristotelian-Scholastic (but then modern) formulation, this meant that the truths of the faith should be taught in the way we learn—that is, through the senses, sight chief among them. Painting founded on the elementally optical, on the elementally sightlike, might "hold the eye" for new purposes, strengthening both instruction and devotion.

Such arguments could justify optical naturalism in painting and theater and the proper uses of illusion in general, which brings us back to Filippo Brunelleschi, as famous in his own time for his religious spectacles[12] and for persuading a gullible neighbor that his life was a dream (a marginally improper use of illusion)[13] as he was for the accomplishments by which he is better known to us. It will be useful briefly to reacquaint ourselves with Brunelleschi and his accomplishments. His name can be made out beneath the unflattering and damaged profile portrait from a panel showing five founders of Florentine art, including Giotto, Donatello, and Paolo Uccello (figure 2.3). The painting seems to be a collection of earlier portraits made in the early sixteenth century. The strict profile of Brunelleschi's portrait suggests that the original was made fairly early in the fifteenth century.[14] In any case, it is an appropriately obscured record of the living appearance of a man who was notoriously secretive and wrote almost nothing.

Brunelleschi was born in 1377, the son of a notary and therefore of higher social station than most Florentine Renaissance artists. He was trained as a goldsmith—there was a lot of gold in Florence in those days[15]—and he entered the stream of the history of art on a somewhat sour note, as a loser in the famous competition of 1401 for the bronze doors of the Florentine Baptistery, a competition won by the young sculptor who would prove to be his lifelong rival, Lorenzo Ghiberti. After this disappointment, Brunelleschi devoted himself to architecture, and, according to his biographer, Antonio Manetti, went with his friend, the sculptor Donatello, to Rome, where Brunelleschi measured ancient buildings according to a method of his own devising.[16] Vasari, writing more than a century after Brunelleschi's death, tells us he had a tenacious mem-

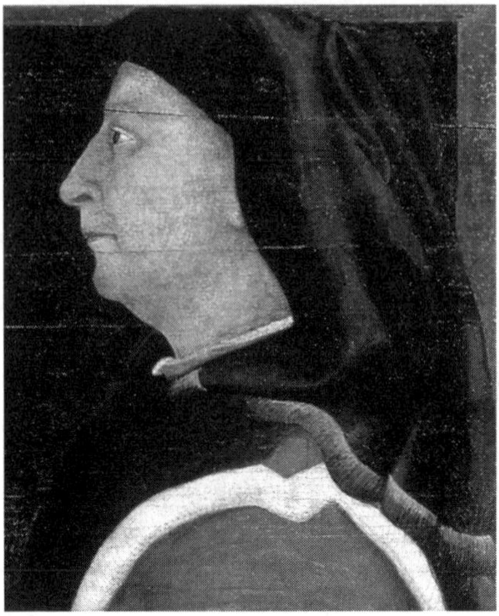

FIGURE 2.3. Five Masters of Florentine Painting, *detail, profile portrait of Filippo Brunelleschi, Florentine School, early sixteenth century, Louvre, Paris. Reunion des Musées Nationaux/Art Resource, New York. Photo: Daniel Arnaudet.*

ory and discoursed with the learned on scripture and on Dante.[17] Brunelleschi stands near the beginning of the Florentine tradition of the *omo sanza lettere* (man without letters), as Leonardo da Vinci called himself,[18] the inventor and discoverer, reliant on experience, associated with the *volgare* and with the marvelous effects of the mechanical arts, a historical figure at least as important to the new period and to unfolding early modernity as scholastic, humanist, or courtier.

Brunelleschi's next major appearance in Florentine art history was a great triumph. He designed and supervised the construction of the dome of Florence Cathedral without using elaborate supporting scaffolding, devising ingenious new machines to accomplish the feat (figure 2.4). Alberti praised this accomplishment in the dedication to Brunelleschi of the Tuscan version of his treatise *On Painting.* Alberti had long believed with others, he says, that Nature had grown old and weary and was no longer able to bring forth either giants or persons of great intellect. But when the ban on his family was lifted (I will treat this more fully in the next chapter) and he returned to Florence, Alberti found all the heroes of the early Florentine Renaissance whose names we still remember, especially Brunelleschi. "Who could be so hard of heart or so envious as not to praise Filippo the architect when he sees here such an enormous construction towering above the skies, vast enough to cover the entire Tuscan population with its shadow? . . . Surely a feat of human ingenuity . . . that people did not consider possible these days and was probably equally unknown and unimaginable among the ancients."[19]

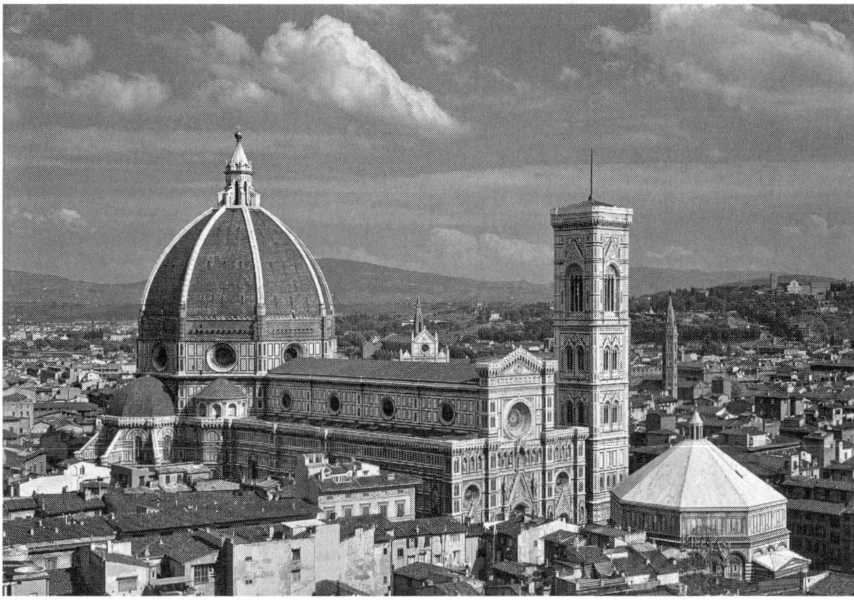

FIGURE 2.4. *Filippo Brunelleschi, dome of Florence Cathedral, 1420–36. Scala/Art Resource, New York.*

Together with building the first great dome in the West since classical antiquity, Brunelleschi formulated a delicate, local classicism, rejecting the foreign Gothic style and establishing himself as the founder of the tradition of European neoclassicism continuing (with fits and starts) to the present. It must be admitted that all of Brunelleschi's enterprises were not as successful, and his record as a military engineer, for example, was spotty, to say the least. Brunelleschi, whose death mask we see in figure 2.5, died in 1446 at the age of sixty-nine. He was buried in Florence Cathedral, where a monument was placed with an inscription hailing him as a modern Daedalus, the legendary Greek builder of the Cretan labyrinth and maker of statues that moved, who might be called the patron hero of the mechanical arts.[20]

And as I have said, Brunelleschi is also credited with the invention of painter's perspective. The date of his first demonstration is uncertain. A letter of 1413 from a poet named Domenico da Prato to a friend in Florence compares life in a country castle unfavorably to the intellectual ferment of the city, one of the attractions of which was the ingenious Brunelleschi—then about thirty-five years old—who is called "prospettivo."[21] All this really tells us, however, is that by 1413, Brunelleschi was considered an expert on optics (*perspectiva*) and that his disquisitions and demonstrations on the subject were causing interest and

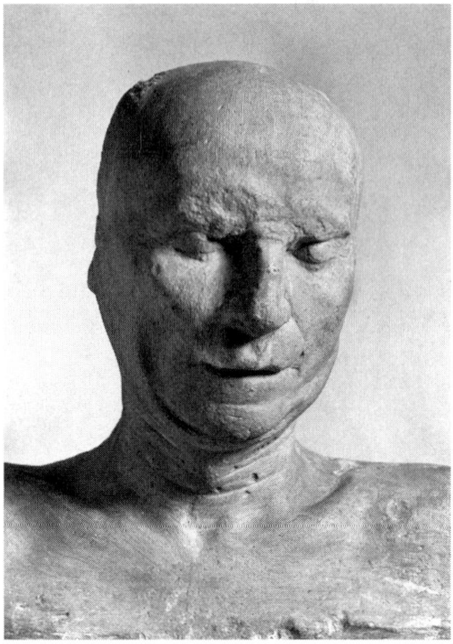

FIGURE 2.5. Death Mask of Filippo Brunelleschi, *1446, Museo dell'Opera del Duomo, Florence. Scala/Art Resource, New York.*

even excitement. As we shall soon see in more detail, optics included all kinds of reflection and refraction as well as direct vision, and so Brunelleschi might have been doing tricks of many kinds in 1413. It is entirely possible that he had devised his version of painter's perspective years before.

Antonio Manetti's biography of Brunelleschi in fact placed perspective among his first accomplishments.[22] Manetti recorded the events of Brunelleschi's life, but he also clearly meant to defend and advance Brunelleschi's reputation, and he may have insisted on Brunelleschi's priority in the invention of painter's perspective by placing it early in his life. But the circumstances Manetti described are plausible, and if we think of Manetti's biography as a chronicle, he set both Brunelleschi's early architectural activities and his invention of perspective between reliefs in Pistoia, associated with documents of 1399, and the competition for the bronze doors of the Baptistery of 1401. Manetti writes that Brunelleschi's architectural aptitudes were recognized early on and that he was commissioned to design and build the quarters of the Monte (the state bank) in the Palazzo della Signoria in Florence. All trace of these quarters is now lost. It is most significant that he did this in a new style, expressly in opposition to prevailing Gothic taste. It might thus be surmised that Brunelleschi's interest in architecture and perhaps in the revival of classical architecture had awakened by 1400. This is important, suggesting that the young Brunelleschi not only studied the forms of classical architecture but also might have been interested in reinventing classical architectural *drawing*.

If Brunelleschi had formulated the principles of painter's perspective so early, why are there no echoes in painting for some two decades? This question has provided a basis for arguments placing his first demonstration around 1425,[23] when the construction finally achieved monumental form in the architectural theater of Masaccio's great *Trinity* in Santa Maria Novella (figure 2.6), probably done with Brunelleschi's collaboration.[24] But perhaps only some artists were inclined to explore the implications of Brunelleschi's new architectural drawing, and the earliest evidence in art for his first demonstration may be the pictorial relief of his friend Donatello, the little portico to our right in the *St. George Slaying the Dragon* predella on Orsanmichele (figure 2.7), which might echo drawings of a portico like that of the Ospedale degli Innocenti (figure 2.8) rather than any actual building.

Against this background, Manetti's introduction to Brunelleschi's demonstrations warrants close attention. First, Manetti tells us that by the time he wrote (around 1480), painters had come to refer to the construction for which Brunelleschi had provided the core as *prospettiva*, although properly the term referred to the whole science of vision—that is, to optics. Painter's *prospettiva* is the "right placement of the diminutions and increases that appear to human eyes, of things at a distance and near at hand, buildings [*casamenti*], plains [*piani*, perhaps also urban spaces], mountains and landscapes [*paesi*], and in every place the figures and other things with that measure belonging to the distance at which they are seen; and from him was born the rule that is essential to [*la importanza di*] everything concerning [painter's perspective] that has been done from then until now."[25]

Manetti thought that painter's perspective was modern, unknown to the ancients (as Alberti also suggested) or, if it was known to them, that it had been lost or was not understood by his contemporaries in any surviving ancient text. Brunelleschi, however, might be supposed to have thought he understood one crucially important text—namely, Vitruvius's account of *skenographia*. I argued elsewhere that when he made his first panels, Brunelleschi meant to reconstruct *skenographia* as the three-dimensional architectural drawing described by Vitruvius[26] and that in effect he coupled Vitruvius's statement that "a central point . . . established at the apex of the extension of the rays of the eyes, corresponding to lines according to natural reasons"[27] with one of the crucial principles of Alhazen's optics, that every point on one facing surface in light communicates with every point on the other.[28] I will presently discuss Alhazen's principle in

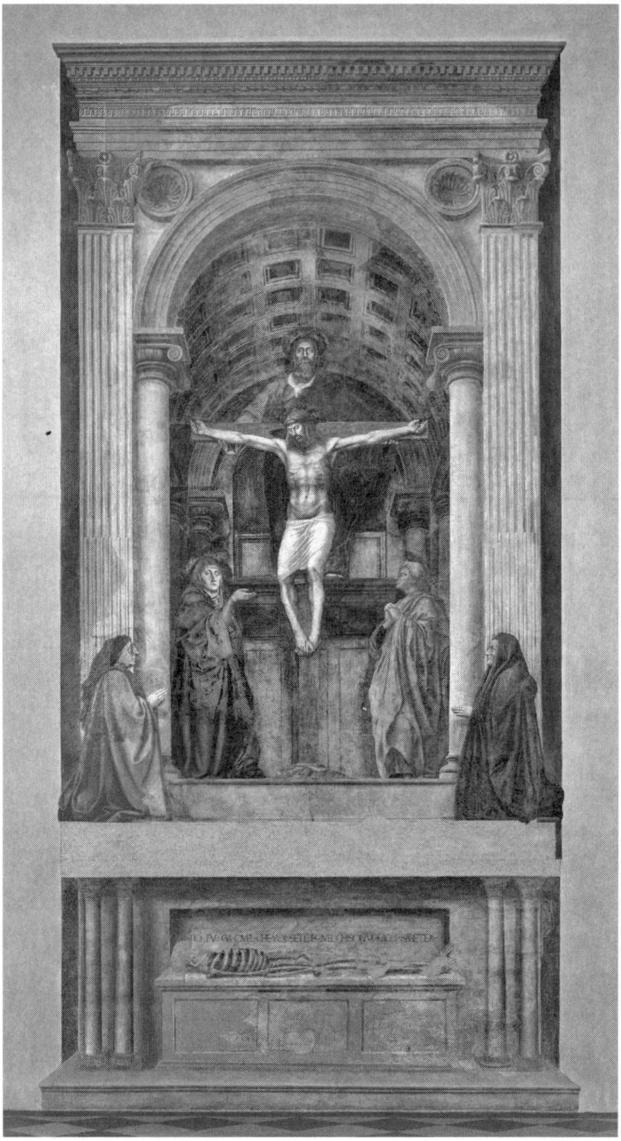

FIGURE 2.6.
Masaccio, Trinity, *fresco, ca. 1425, Santa Maria Novella, Florence. Scala/Art Resource, New York.*

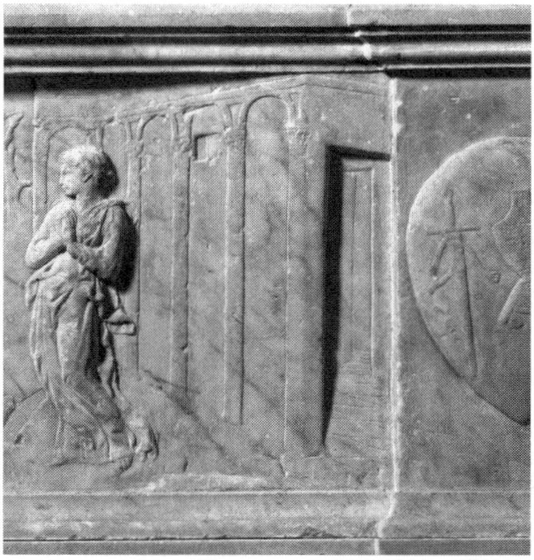

FIGURE 2.7. *Donatello,* St. George Slaying the Dragon, *marble relief, detail, ca. 1417, Orsanmichele, Florence. Scala/Art Resource, New York.*

more detail. In the meantime, Brunelleschi's elegantly simple fusion of the ancient definition of point of view with respect to painting and architecture and the beginnings of the modern theory of light yielded the core of what would become painter's perspective and grounded it in natural principles.

At the end of his successful career, Lorenzo Ghiberti, Brunelleschi's perennial rival, turned his attention to the same Vitruvian text in his *Commentarii*, the first autobiography by an artist. In his first commentary (of three), Ghiberti gave Agatharcus and his lost treatise an important place in the history of ancient art,[29] and in the third, near the end, he placed Vitruvius's text in his discussion of the proportions of the *statua virile*. For the most part, Ghiberti's treatment of *perspectiva* in the third commentary is made up of texts from the medieval optical writers I have been discussing and has nothing at all to do with painter's perspective. His use of Vitruvius, however, shows how easily this authoritative (and available) text could be expanded and adapted to then-current interests. Ghiberti's very approximate translation of Vitruvius turns into an explanation in more modern terms: "if one goes to where Agatharcus has set the eye, in the said work the lines will go as they go in nature, in the *virtù visiva*. And having gathered the true boundaries of the said lines . . . they have figured many things in certain places, showing how strong [*di quanto valore*] is the said art, and how marvelously it fools human nature." Whoever diligently pursues the art of painting, Ghiberti concluded, will follow the example of Agatharcus.[30]

Graziella Federici-Vescovini argued that Brunelleschi's first perspective demonstration, which she placed very cautiously between 1403 and 1425, was a result of what might be called an Alhazen revival in the late fourteenth century.[31] Alhazen's optics offered manifestly superior explanations for both vision

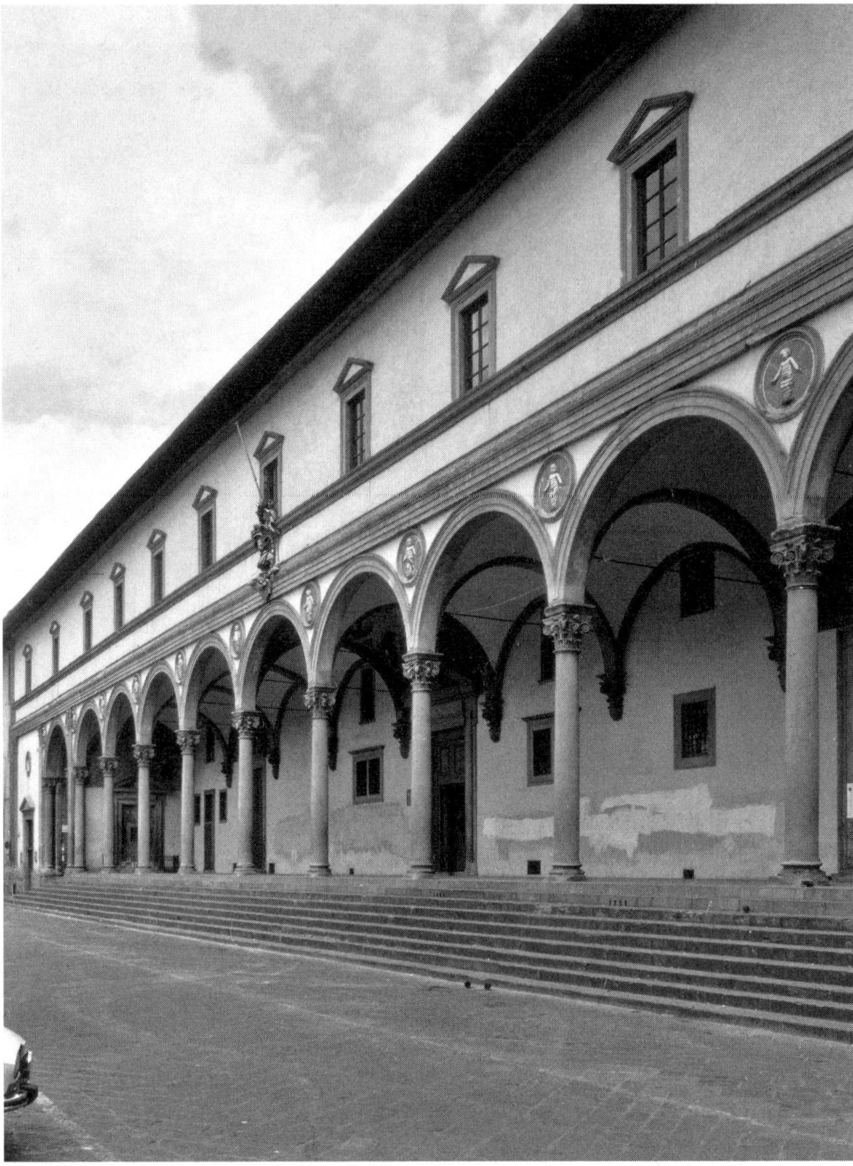

FIGURE 2.8. *Filippo Brunelleschi, loggia of the Ospedale degli Innocenti, Florence, begun 1419. Alinari/Art Resource, New York.*

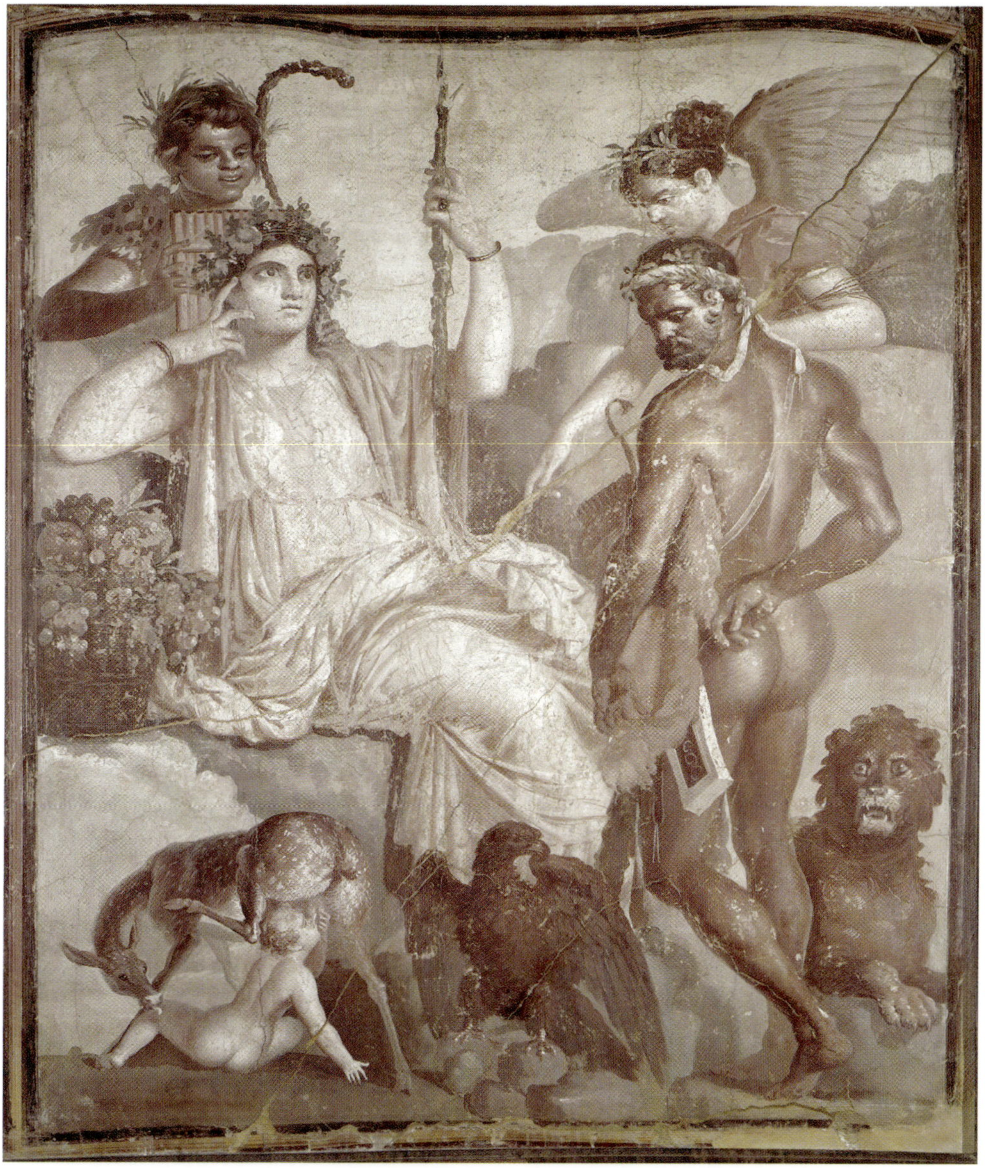

PLATE 1. Herakles Discovering Telephus, *from the Basilica at Herculaneum, Roman copy after Greek original of the fourth century* B.C.E.*, Museo Archeologico Nazionale, Naples. Scala/Art Resource, New York*

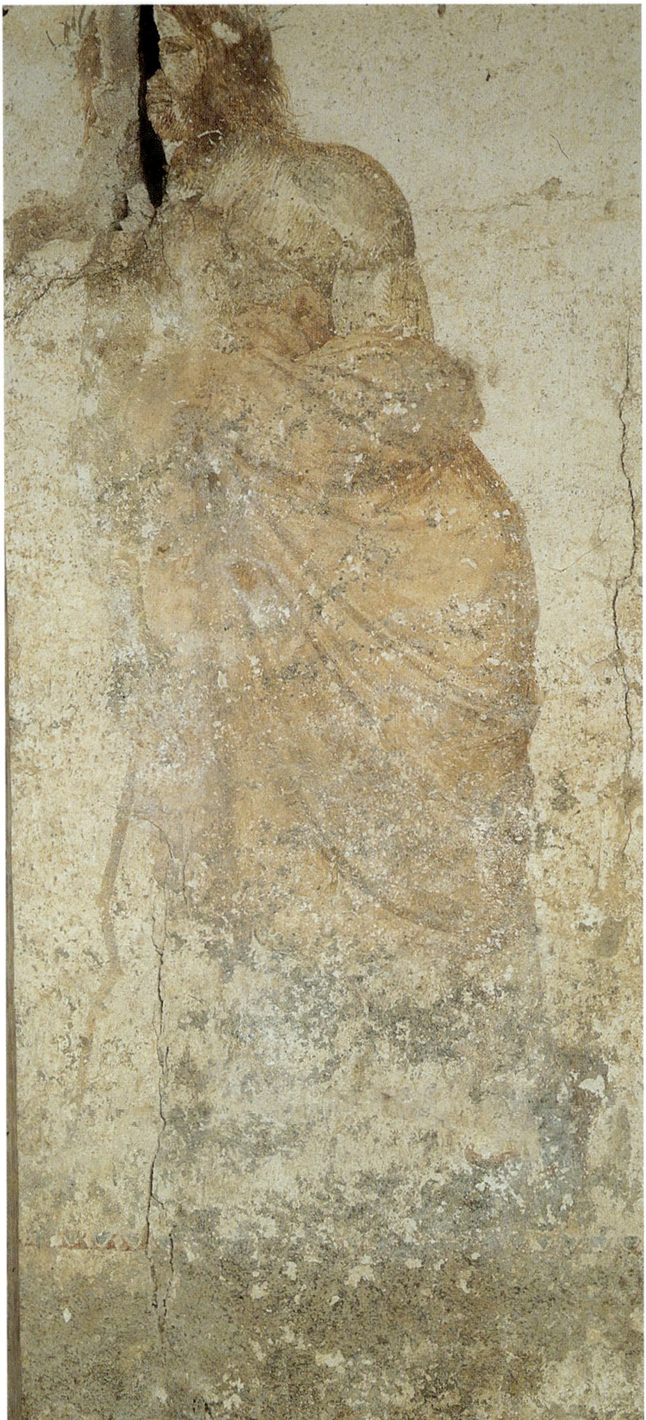

PLATE 2.
Rhadamanthus, *from the Great Tomb at Lefkadia, late fourth century* B.C.E.

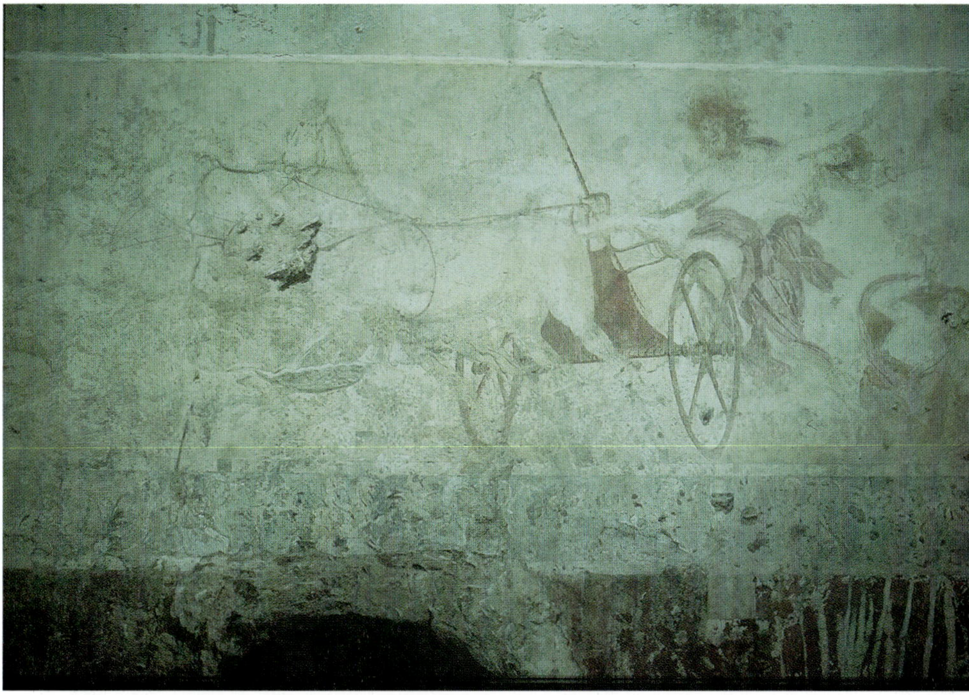

PLATE 3. Abduction of Persephone, *from Tomb 1, Vergina.*
© Archaeological Receipts Fund, Hellenic Republic, Ministry of Culture.

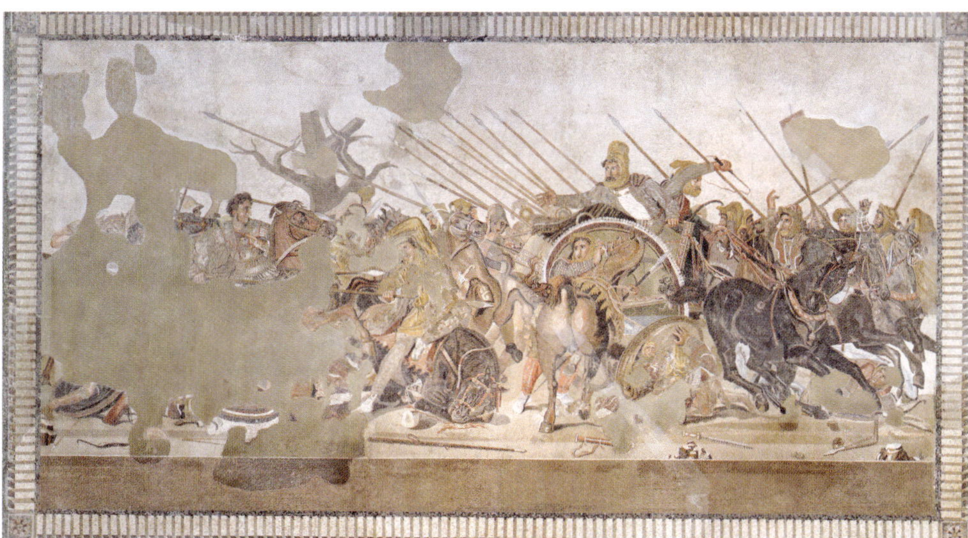

PLATE 4. Battle of Alexander the Great and Darius, *Roman mosaic from the House of the Faun, Pompeii, second–first century* B.C.E., *after a fourth-century original by Apelles(?), Museo Archeologico Nazionale, Naples. Alinari/Art Resource, New York.*

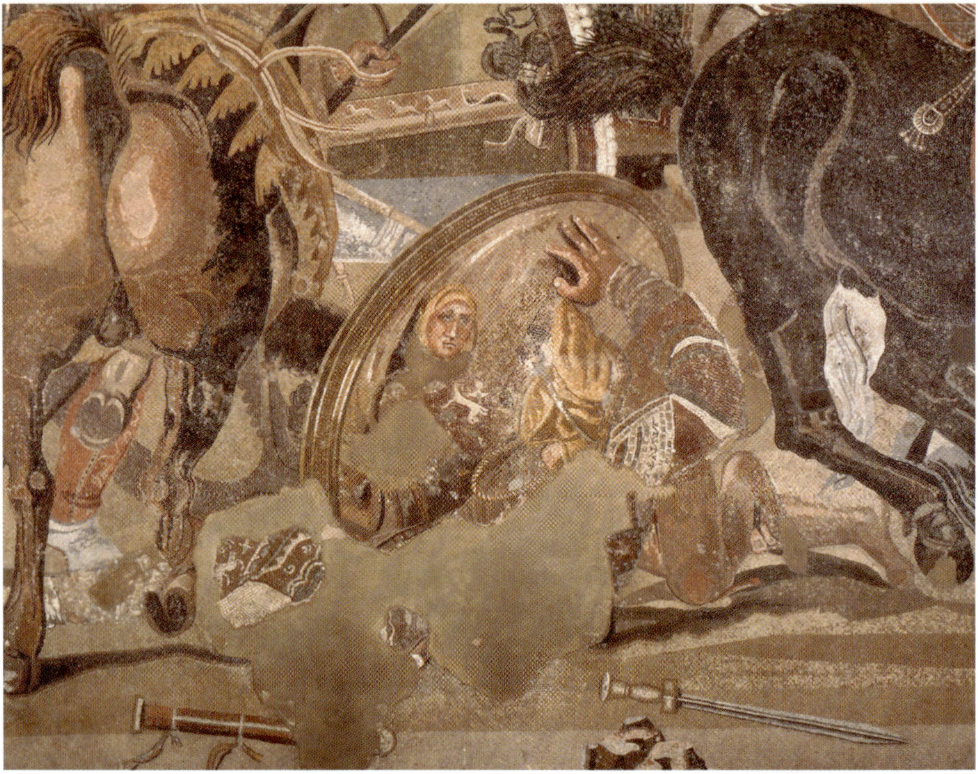

PLATE 5. Battle of Alexander the Great and Darius, *detail of figure 1.12 / plate 4, a fallen warrior and his reflection in a shield.*

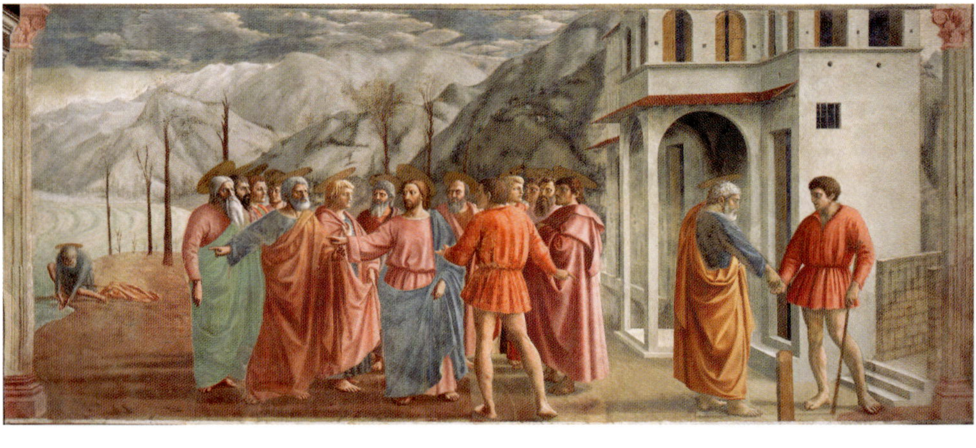

PLATE 6. *Masaccio,* Tribute Money, *fresco, 1420s, Santa Maria del Carmine, Brancacci Chapel, Florence. Scala/Art Resource, New York.*

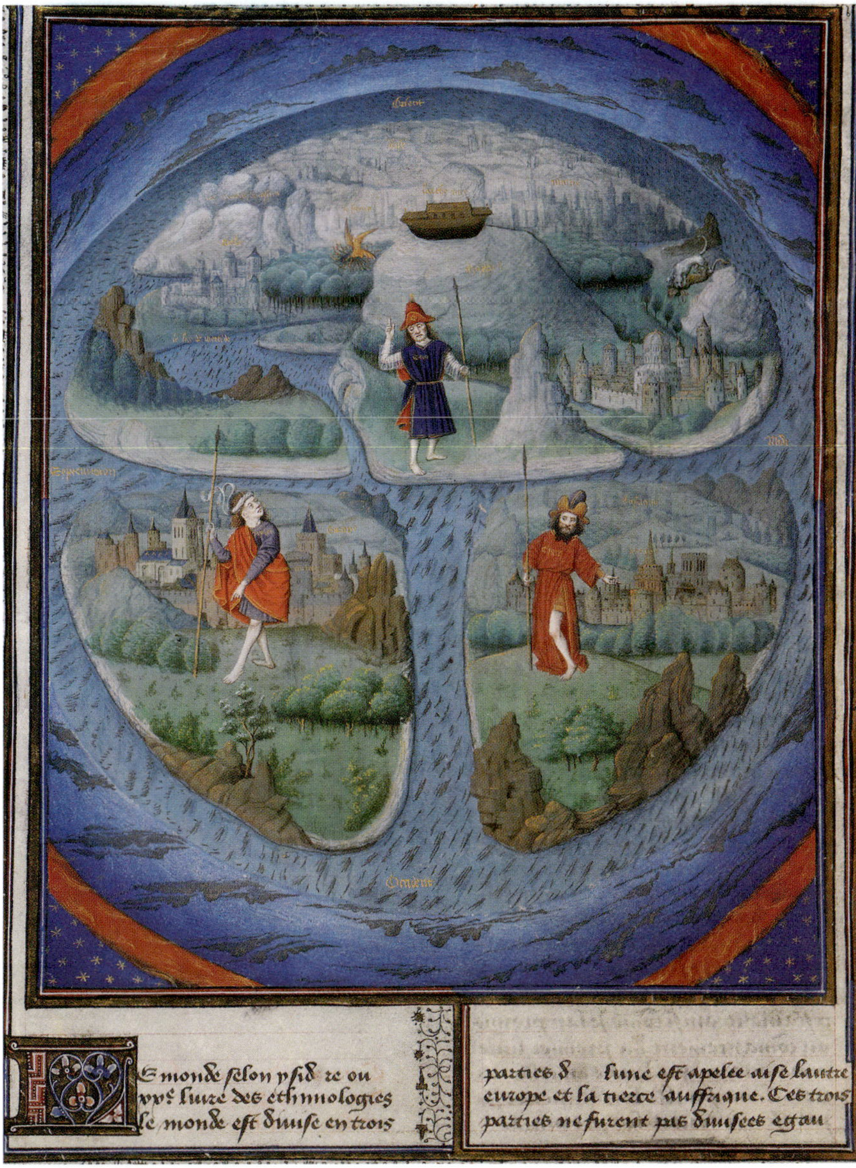

PLATE 7. *Simon Marmion,* World according to Isidore of Seville, with the Sons of Noah. *From Jean Mansel,* Fleur des histoires *(ca. 1460).*

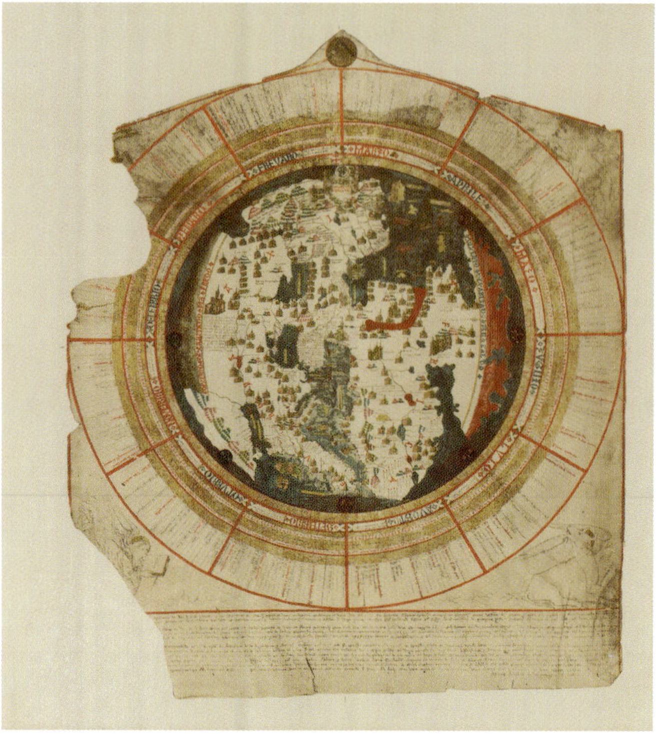

PLATE 8. *Giovanni Leardo,* World Map, *Venice, 1452, American Geographical Society Library, University of Wisconsin–Milwaukee Libraries.*

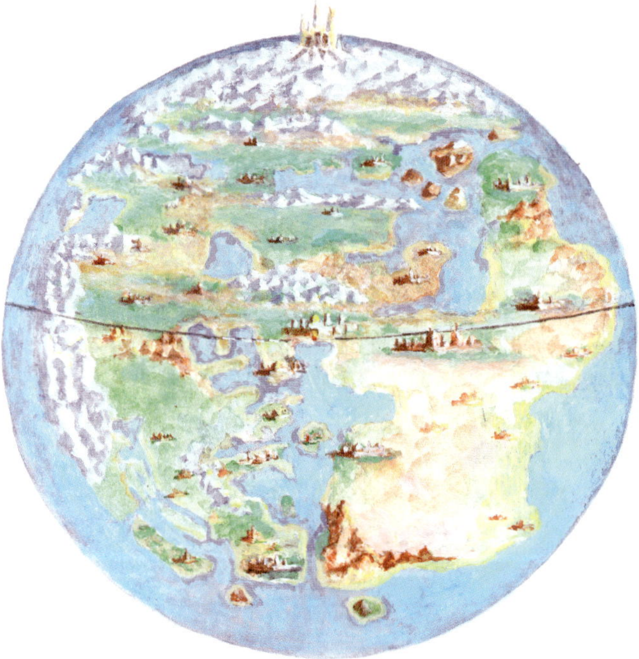

PLATE 9. *Author's suggested reconstruction of Jan van Eyck's lost map. Cities are fewer and not to scale. There was probably a circle of air with clouds and perhaps a circle of fire to complete the four elements.*

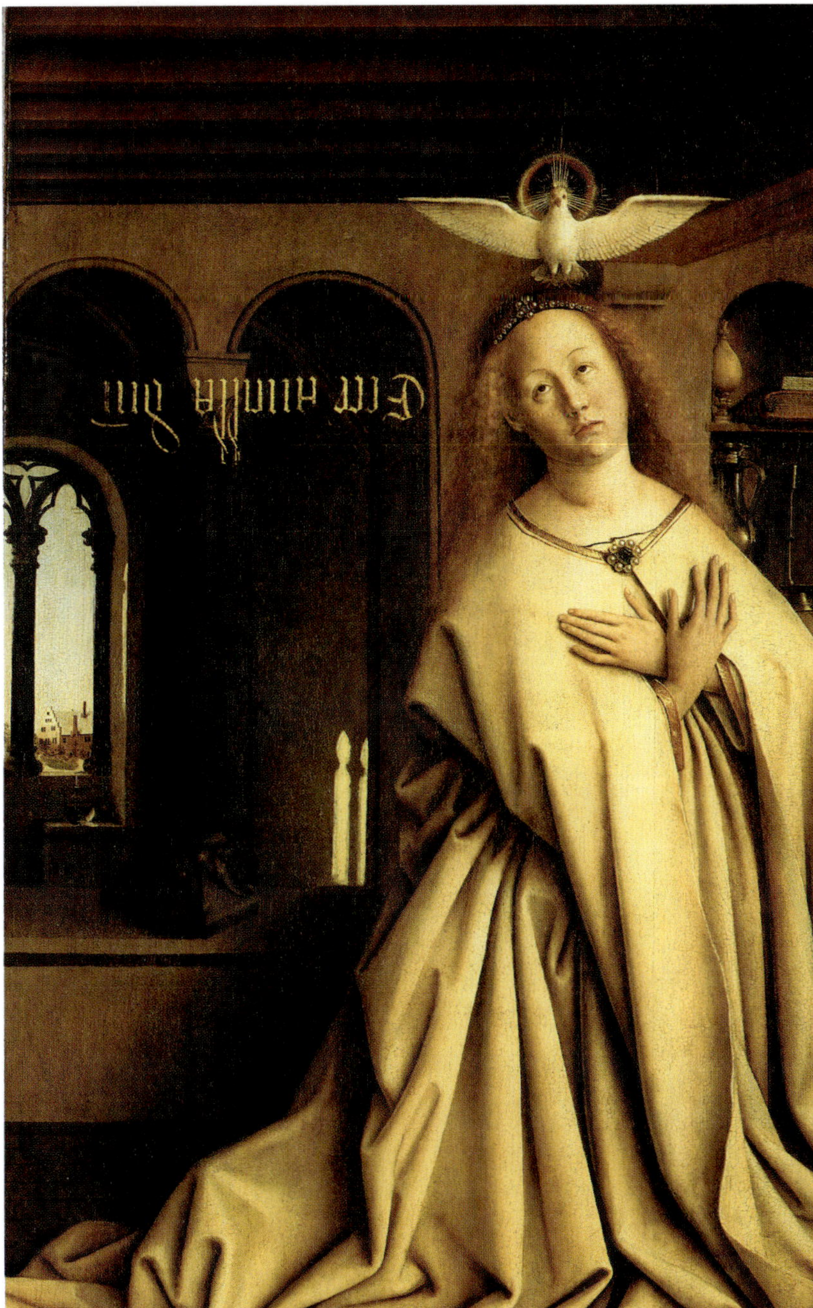

PLATE 10. *Jan van Eyck, Ghent Altarpiece, closed, detail of the* Annunciation, *flask, and lights on the wall behind the Virgin Mary, completed 1432, Cathedral of St. Bavo, Ghent. Scala/Art Resource, New York.*

PLATE 11.
Jan van Eyck,
Madonna and Child
(Lucca Madonna),
oil on panel, ca. 1434–35, Städelsches Kunstinstitut, Frankfurt. Photo © Ursula Edelmann, Artothek.

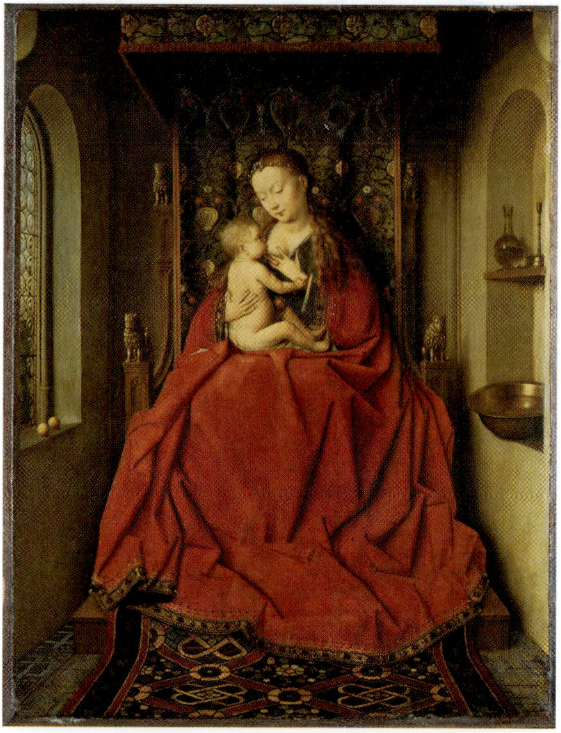

PLATE 12.
Hans Memling, Last
Judgment, *detail of figure 4.20.*

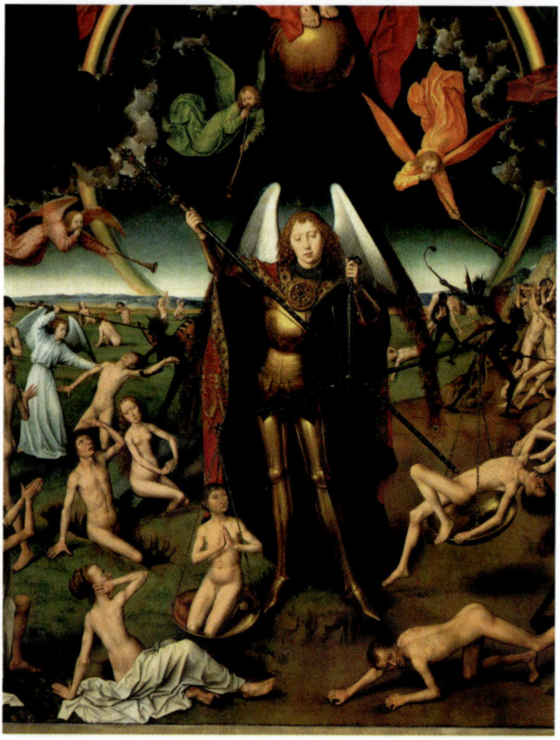

and perception, but it was difficult to reconcile with more familiar Aristotelian and—especially—Platonic principles. This is because it was based on more nearly materialist Stoic ideas, and when Alhazen's optics was translated into Italian in the late fourteenth century, it presented a new alternative to its own intervening adaptations and interpretations. Chief among these new interpreters—still following Federici-Vescovini—was Biagio Pelacani da Parma, who lived from 1347 to 1416. Instead of adapting Alhazen's optics to prevailing Scholastic thought, Pelacani set about to construct a new psychology and a new natural philosophy on Alhazen's principles. In very general terms, he argued that qualities are always quantities, that individual things are unions of active qualities, that sensation is always an interaction of two surfaces, and that a vacuum positively exists in the sense that it is always possible to measure relations between individual existing things.[32] Abstract as these principles may sound, the last two are especially relevant to the present discussion, and I am going to argue that Brunelleschi's first perspective panel demonstrated similarly progressive early modern ideas with a concreteness and clarity not unlike his legendary demonstration that an egg could be made to stand upright by smashing one end of it on a table, a solution sufficiently compelling to demonstrate once and for all that he alone was equal to the task of building the dome of Florence Cathedral.[33]

Biagio Pelacani da Parma taught in Padua but was connected to Florence through his student, Giovanni di Gherardo da Prato, a poet and all-around scholar who lectured in the Studio Fiorentino on Dante—a subject on which, as we have seen, Brunelleschi was also an expert. Giovanni da Prato was Brunelleschi's friend until about 1410 and then his enemy, best remembered for his objections to Brunelleschi's plan for the cathedral dome and for the exchange of splenetic sonnets between the two men in which they insulted one another as "*animali*."[34] Giovanni da Prato was entrusted by the Studio Fiorentino with recruiting visiting faculty and brought his old teacher, Pelacani, from Padua to Florence in the summer and fall of 1388 to lecture on moral and natural philosophy. We know from the date of its earliest manuscript that Pelacani had written his *Quaestiones de perspectiva*, a commentary on the *Perspectiva communis* of John Pecham, by 1390.[35] Giovanni da Prato's *Paradiso degli Alberti* was probably completed in the 1420s, but Pelacani was cast as an interlocutor in the late 1380s and is described as learned and specifically as learned in optics.[36] This example urges the conclusion that Brunelleschi's solution that would change the art of painting and visual representation to the present arose, like Florentine humanism, around groups of individuals and from long, interested conversations and disputes of which we now have only traces and fragments.

The perspective construction codified by Alberti in his treatise of 1435 must also have been the fruit of ongoing conversations and rivalries, friendly and otherwise. James Ackerman stressed the scientific interests of Florentine humanists in the early fifteenth century and associated Alberti with the circle of Niccolò Niccoli and Ambrogio Traversari. The members of this circle were supporters of Cosimo de' Medici, opposed to the older oligarchy that had banished the Alberti[37] (as I will discuss in the next chapter). According to his biography by Vespasiano da Bisticci, Niccolò Niccoli was a close friend of Brunelleschi, Donatello, Luca della Robbia, and Lorenzo Ghiberti.[38] It may be noted that only Masaccio is missing from the four artists listed in Alberti's dedication to Brunelleschi of his treatise *On Painting*.

In such circumstances, Brunelleschi might have had access to the best philological assistance, and the project of reconstructing Roman architectural drawing as described by Vitruvius as well as reclaiming the orders corresponds closely to the recovery of Roman orthography and grammar. In any case, Brunelleschi might have provided the basic contribution around which new refinements, clarifications, and practical solutions continued to be made. Brunelleschi's first demonstration of the reinvention of ancient architectural drawing in light of then-modern optics was based on three assumptions: First, a fixed point of view (although the "point" in question was physical and therefore a surface); second, that vision is the interaction of surfaces in light, one of which—the "point" of view—is the sensitive surface of the eye; and third, the comparison of the sensitive surface of the eye to a plane mirror, in which a true reflection is possible.

Optics was an early experimental science, and Brunelleschi's demonstration may be supposed to have been provisional and hypothetical. His construction yielded diagonal lines converging on a line of sight rather than beginning from a point, and his construction was neither explicitly anthropometric nor modular. It did, however, require measured drawings scaled to *braccia*, pointing to a deep and essential difference between ancient *skenographia* and Renaissance painter's perspective, which (after Alberti) incorporates ratio from the beginning and is therefore mathematical as well as geometric.

Insistence on the fairly gradual separation of a distinct "painter's perspective" from more general optics is of primary historical importance because it points to quite different precedents and chains of evidence than does the assumption either of invention or discovery, and it is of fundamental historical significance that the plane mirror provides the example for the first crucial unifying pictorial surface, the principle of coherence for what Alberti would call the "intersection," his "open window," framing and cutting across the many optical pyramids connecting objects to the eye.

Like Vitruvius and even as a learned artist himself, Alberti claimed to have no interest in the question of whether visual rays issue from the eye (the old Platonic extramissive view) or enter it (the modern intromissive view of Avicenna and Alhazen) because the geometry was the same in either case.[39] Intricate arguments about the geometry of rays inside the eye could simply be bypassed in favor of the geometry of the external world in light. Alberti also dismissed the problem of where exactly the act of vision occurs, although his words make it clear he understood the alternatives. Near the beginning of the Latin version of his treatise on painting, Alberti observed that "this is not the place for disputing whether vision might rest in the junction of the interior nerve of the eye, as some wish, or whether images are figured on the surface of the eye, as if in a living mirror [*quasi in speculo animato*]."[40] In order to understand Alberti—and perhaps to reclaim something more of the conversations I have just imagined—it is necessary to turn to a basic part of optics I have not yet had occasion to consider, namely, the physiology of the eye. Medieval understanding of the eye's anatomy was based on the second-century physiology of Galen, who located the crucial sensitive surface in the crystalline humor. We now consider this a lens, and the retina, the "little net" at the back of the eye, to be what medieval writers called the "seat of vision," but we owe this modern correction of the optical tradition to Johannes Kepler, who, following closely on advances in anatomy through the sixteenth century, published his *Ad Vitellionem paralipomena* (*Additions to Witelo*) in 1604.[41] In general, the eye was considered to be the organ (the tool) of the visual virtue—that is, the power of the human soul to see—and the crystalline humor was the unique site of the capacity of this virtue, to which, in the order of things, the other parts of the eye were subordinate.

The shape of the crystalline humor itself will be a recurrent theme in what follows. Galen wrote that it was not perfectly spherical (even though Nature loves spheres) so that it might nest more stably in the spheres surrounding it.[42] Galen, of course, was an extramissionist, and the intromissionist Alhazen saw quite different implications in the same configurations of the eye. Alhazen gave a new level of microcosmic clarity to the eye's spheres (figure 2.9), the centers of which all coincide with the axis connecting the center of the pupil with the center of the optic nerve. All rays admitted to the eye meet (in principle)[43] at the center of the sphere of the eye itself.

The crucial crystalline humor was made up of two parts, the glacial (ice-

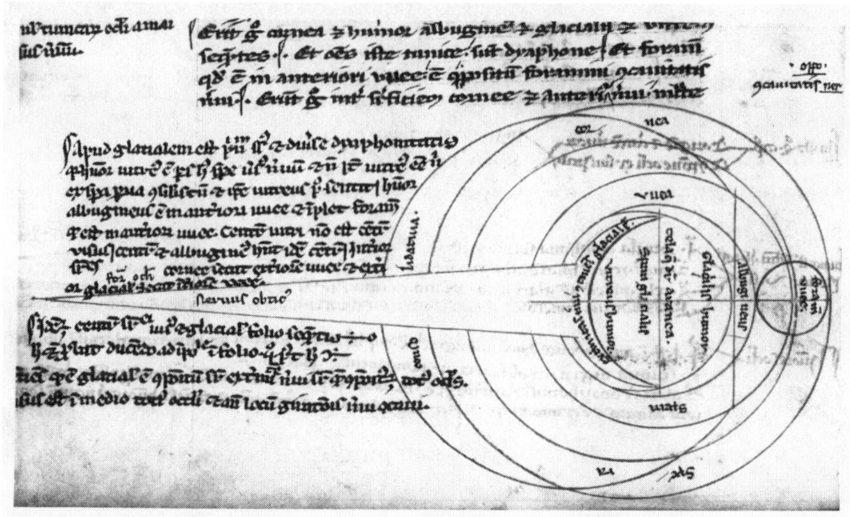

FIGURE 2.9. *Diagram of the eye from a thirteenth-century manuscript of Alhazen,* De aspectibus. *Crawford Library, Cr.3.3., Royal Observatory, Edinburgh.*

like) humor and the posterior vitreous (glasslike) humor. In figure 2.9, from a thirteenth-century manuscript of a Latin Alhazen, the two parts form a mandorla shape, sharing a common border, a straight line perpendicular to the axis of the spheres of the eye; this mandorla shape is determined by sections of circles of diameter larger than that implied by this line. Alhazen himself described the crystalline humor as a "white, small, humid sphere," diaphanous and therefore receptive to forms, because of its essentially watery nature, adding that there is a small compression on the surface of the front part of this sphere, "like the compression of the surface of a small lentil [*lenticula*]."[44] He also described this compression as a section of the arc of a larger circle (hence its mandorla shape in figure 2.9) and said that it responds to the shape of the pupil, which admits light to the eye. This may be taken to mean that light rays pass more nearly in parallel into the eye and to the crystalline humor, preserving the same pattern of light, which Alhazen insists again and again is essential. The glacial and vitreous humors then become the means by which this minimally distorted pattern is refracted and then re-refracted to the optic nerve.

Alhazen continually restates and rewords his premises as his arguments build. He worked from a very inadequate base of anatomical knowledge, and much of his reasoning is essentially geometric. Given the structure of the eye as he understood it, what must be the case if the pattern of light corresponding to the field of vision is reliably transmitted to the optic nerve and to the

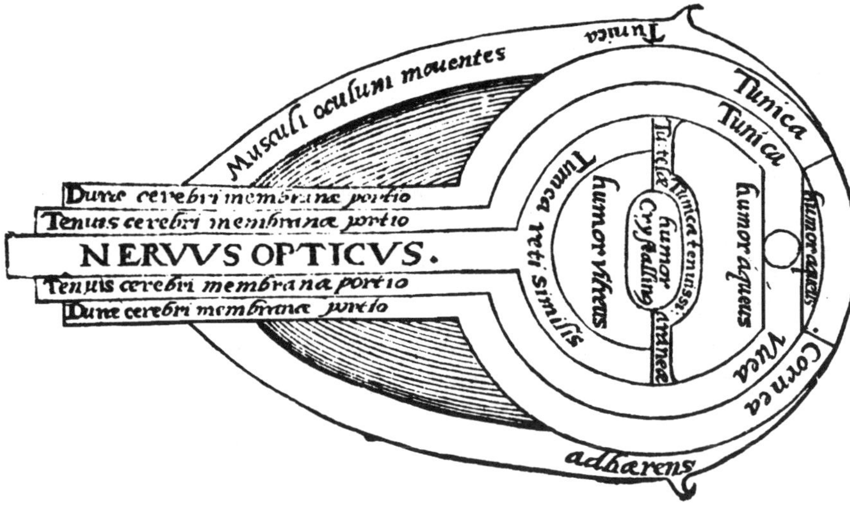

FIGURE 2.10. *Diagram of the eye from Risner's* Opticae thesaurus, *6.*

postsensationary faculties where the judgments are finally made that complete vision? As all of this might suggest, Alhazen's text was differently understood and illustrated, as a comparison of figures 2.9 and 2.10 will show. In figure 2.10, from Risner's *Opticae thesaurus*, the crystalline humor is shown in the shape of a pill. What is shared, however, is the representation of the seat of vision as a more nearly planar surface.

An important tradition of commentary descends from Aristotle's *De sensu et sensato*, part of the so-called *Parva naturalia*, the cluster of essays supplementary to his *De anima*, the foundation of the Western science of psychology. Aristotle rejected the position of Democritus, who argued that the act of vision is the reflection to be seen on the surface of the eye (as we still speak of the "pupil" of the eye, from the tiny reflected image (Latin *pupilla*, equivalent to the Greek *kore*, "maiden"). If the act of vision takes place on the surface of the eye, Aristotle objected, why did Nature, who does nothing in vain, make eyes so inwardly intricate? The eye, Aristotle concludes, simply reflects because it is very smooth, and Democritus could therefore not explain why any shiny object is not an eye. Aristotle did not simply reject the idea that the eye was something like a reflective surface, however. Democritus, he said, was right to think that the eye is essentially watery, but the sensitive surface in question lies within the eye, not on its outer surface. Aristotle did not know what this inner surface might be or how it might relate to other anatomical complexities, and, as we have just seen, the millennial clarification of the physiology of optics moved the crucial sensitive surface more and more deeply into the eye, to the crystalline humor and finally to the retina.[45]

Visible objects either give off light, like the sun, candles, and lightbulbs, or, like most of the things we see, *reflect* light. When Alhazen took up the subject of mirrors and reflection in book 4 of his *Optics*, he began by referring the reader back to the principles of book 1: "light from a lighted body, whether proper or accidental [that is, whether the body is a source of light or simply *in* light] is directed to every body opposite to it." The same is true of color.[46] That is, on a bright day, the sun's diffused and endlessly scattered light makes contact, from a distance and through a medium, between countless facing colored surfaces and the surfaces of the eyes that see them.

The metaphor of the eye as a mirror assumed new importance with the resurgence and final triumph of Aristotle's intromissive theory of vision. The eye was not itself a source of light, as writers from Plato to Euclid to Galen had argued, it was instead passive, the receptor of light, behind and beyond which other faculties come into play. The eye, Avicenna wrote, is like a mirror, and "if a mirror should possess a soul, it would see the image that is formed on it."[47] For Averroes—the Commentator, as he was called by the Western Scholastics—the common sense, the particular intellect, lies behind the retina.[48] The Dominican protoscientist Albertus Magnus called the eye a "*speculum animatum*" in his *De homine* and in his commentary on Aristotle's *De anima* called the eye a "*speculum animatum convexum* [a living convex mirror]." The problem posed by the doubleness of the eye, he said, had been solved by Avicenna, who argued that the two images are unified at the junction of the optic nerve—Alberti's other alternative—which Albertus compared to the apex of a triangle.[49]

I have already mentioned that the vision of Dante's *Divina commedia*—in addition to its purely poetic attainments—was a summa of medieval learning and so was expanded and illuminated by the new optics. In his *Convivio*, Dante described the process of vision in terms largely set by Aristotle's *De sensu*. Perhaps his reliance on Aristotle put Dante ahead of his time (and in line with the new optics) in his firm rejection of extramission. He characterized the eye in words Aristotle had used for the soul: it is not the thing but rather the form of the thing[50] that passes through the medium (space, air), which Dante compared to transparent glass. The "water" of the "pupil of the eye" is also transparent, but the visible forms do not continue indefinitely as they do through the medium because the pupil (by which he must have meant the reflective surface of the eye, perhaps the crystalline humor) is like a lead-backed mirror, and therefore forms appear in it. Perhaps he had in mind the anterior and posterior parts of the crystalline humor. Whether or not that is so, the visual spirit carries the form from the "pupil" to the front part of the brain, where the sensitive virtue has its origin and font (presumably the common sense). This all happens instantaneously, and so, Dante concluded, we see.[51]

As I noted in passing, Alberti's words—"*speculum animatum*"—appear in the Latin version of his treatise on painting and not in the Tuscan version, a copy of which he dedicated to Brunelleschi. Although debate continues, consensus seems to be emerging that the Latin was written first.[52] In any case, it is generally recognized that the two versions were meant for different audiences and purposes—courtly and liberal in the first case, craftsmanly and mechanical in the second. But perhaps the reason for the difference between the two texts was even more closely circumstantial and personal: Alberti entirely omitted any mention of the issue of the seat of vision from the Tuscan version because Brunelleschi himself defended the idea that the eye is a living mirror, while Alberti, by the time he wrote his treatise, had adopted another metaphor, with different implications. The eye, agent of the *virtù visiva*, is not the mirror of the world, it is the window[53] of the soul, through which the soul surveys the world.

This brings us at last to Brunelleschi's first perspective demonstration, which may have been both serious and playful, as experiments—especially those involving illusions—can be. It took place on familiar art-historical ground, in the central portal of Florence Cathedral (figures 2.4 and 2.11). Figure 2.11 is a late-sixteenth-century drawing of the early-fifteenth-century facade of the cathedral, and we may imagine Brunelleschi standing in the central doorway. Depending on the year, the figures of the four Evangelists, among them Donatello's *St. John the Evangelist*, its proportions adjusted *di sotto in su*, may have been in the flanking niches.

Brunelleschi's demonstrations are described in detail by his biographer, Antonio Manetti, who wrote at least sixty years after the events but also claimed to have held the first panel in his hands.[54] This panel was square, half a *braccio* on a side—that is, just under twelve inches—and showed the eastern face of the Florentine Baptistery (figures 2.4 and 2.12). Brunelleschi stood, Manetti reckoned, some three *braccia* inside the cathedral portal, and, looking out almost directly at the Baptistery, showed "as much as one sees at a glance to the sides looking out."[55] The Baptistery was fundamental to the Florentines' sense of their classical past and to Brunelleschi's own forming neoclassical architectural vocabulary, and he may be supposed to have measured it and studied it closely. And, as we have seen, the same interest in architecture might be expected to have led him to Vitruvius's description of the architectural drawings of the ancients, of buildings in plan, elevation, and the three-dimensional projections of *skenographia*.[56]

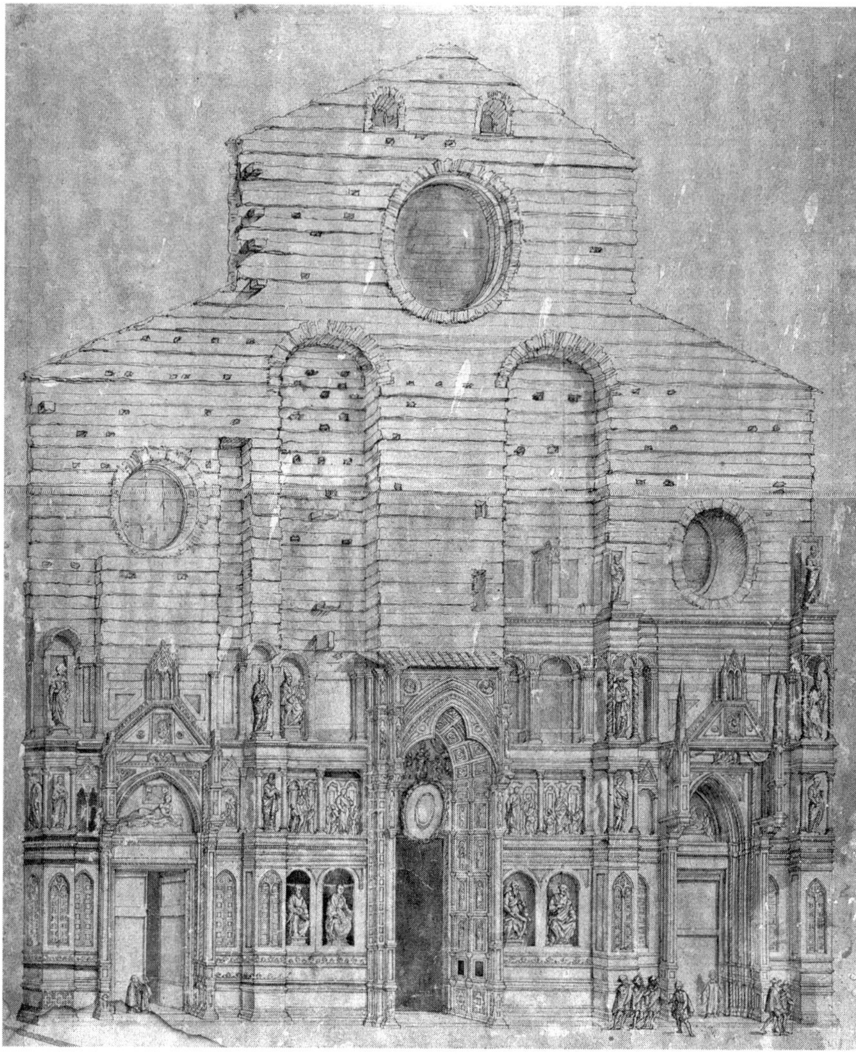

FIGURE 2.11. *Bernardino Poccetti,* Old Facade of Florence Cathedral, *before 1588, Museo dell'Opera del Duomo, Florence. Scala/Art Resource, New York.*

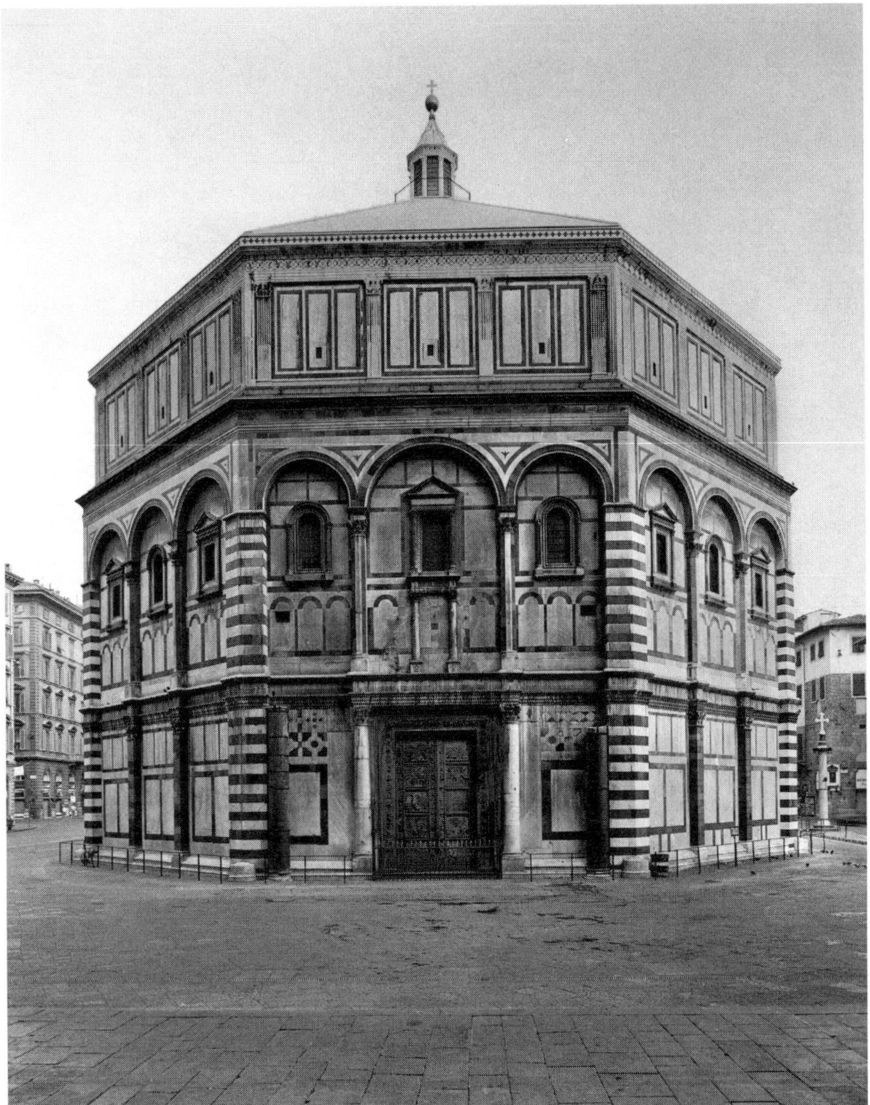

FIGURE 2.12. Baptistery of San Giovanni, *ca. 1060, Florence. Photo: Ralph Lieberman.*

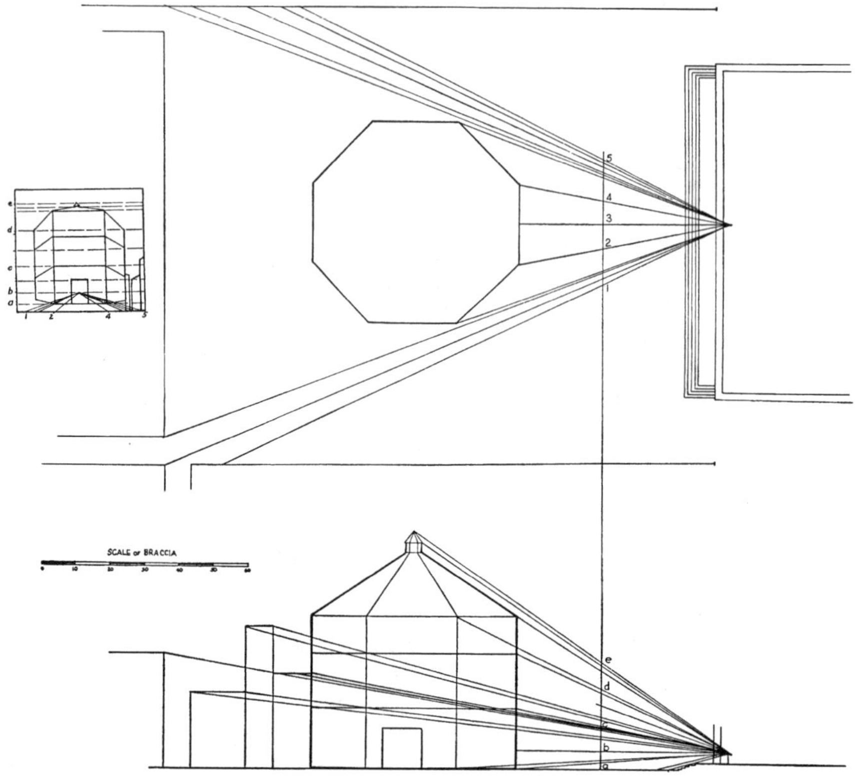

FIGURE 2.13. *Plan, side elevation, and construction of the Baptistery in the Piazza del Duomo. From Krautheimer and Krautheimer-Hess,* Lorenzo Ghiberti, *1:239, diagram 4.*

I have already said that I do not think Brunelleschi invented painter's perspective all at once, and I am persuaded by Richard Krautheimer and Trudy Krautheimer-Hess's explanation, following Vasari, that Brunelleschi used scale drawings of the plan and side elevation of at least the facade of the cathedral, together with the facing piazza and Baptistery.[57] Plan and elevation of the whole piazza would have been necessary because Manetti gave some indication of the visual angle assumed by identifying the buildings flanking on either side (figure 2.13). With the necessary drawings in hand, Brunelleschi would have identified his own position in plan as the apex of the angle in the central doorway of the cathedral and established his line of sight (3) as a perpendicular from that apex to the doorway of the Baptistery. In side elevation, he would have established his scale height (b) at the same point, the apex of the angle in elevation. From the two apices defining his location, he drew visual angles to salient features of the symmetrical Baptistery and buildings flanking the square. These points—

12345 in plan and *abcde* in elevation—were then marked at the same distance, where the base of the visual angle was equal in plan and elevation to the breadth and height of the square panel. These points were extended as horizontal and vertical lines to form a skeletal grid of the Baptistery. Finally, Brunelleschi connected the dots and filled in the blanks, in Manetti's words, as well as any miniaturist might have done.[58]

So far, Brunelleschi's panel has yielded the core of one-point perspective construction. This is complicated, however, by the fact that the sky above the piazza was not painted; rather, that part of the panel was silvered, thus becoming a mirror. As Manetti wrote, "and insofar as he had to show the heavens, that is, the silhouette of the buildings [*muraglie*] in the painting against the air, he put burnished silver, so that the air and natural heavens are seen reflected there, and also the clouds that are seen in that silver to be stirred by the wind when it blows."[59]

It is important to bear in mind that Brunelleschi's first panel was not just a painting, it was a demonstration (figure 2.14). Moreover, it was not simply to be looked at, it was to be seen under very specific conditions. A conical hole was drilled in the panel at the point corresponding to the line of sight. This hole was the size of a "lentil" on the front, opening to the circumference of a coin in the back. The viewer pressed an eye against the opening in the back of the panel and looked through it to see the reflection of the *front* of the panel in a small plane mirror held at arm's length. (Since the angle of incidence equals the angle of reflection, the small handheld mirror need only have been a quarter the area of the panel, about six inches square.) Through the opening, the painted Baptistery would have been seen as a soft-edged circle to which the sides of the small square mirror were roughly tangent. (The handheld mirror might therefore also have been round, six inches in diameter.)

Why this indirect, reflected viewing of a composite painted and mirrored surface, the whole of which was only to be seen as a reflection? What was Brunelleschi demonstrating?

If, as I have suggested above, Brunelleschi meant to reclaim the architectural drawing Vitruvius called *skenographia*, and if, as I have also suggested, Brunelleschi applied Alhazen's principle of the communication of points on surfaces in light to do so, he would have demonstrated this communication simply by having shown that from a point (or opening, a set of points, a surface) on the painted panel, all of that panel's surface could be seen on the facing surface of the handheld mirror. Not incidentally, Alhazen's principle was also an early but very clear statement of the modern theory of light, which Leonardo da Vinci would later call nothing less than the "first principle of the science of painting,"

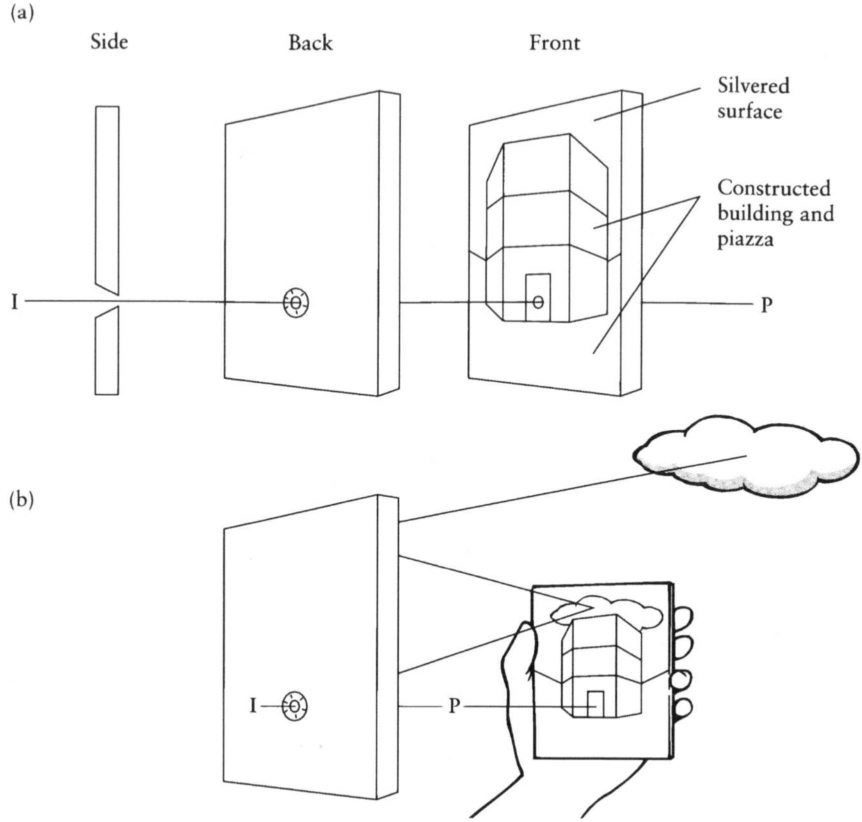

FIGURE 2.14. *Brunelleschi's panel in use. From Summers,* Real Spaces, *fig. 289.*

in effect identifying painting with optics.[60] So far, Brunelleschi's demonstration shows that painted panel, mirror reflection, and eye are equivalent, but this leads to two further questions: Why was the sky of the panel mirrored? And why were the constructed virtual space and the mirrored surface of the panel treated as one continuous surface?

More must be said about mirrors. Metal-backed mirrors began to be made in the late Middle Ages. Most were convex (or concave) because fairly large regular spheres of glass could be blown and sections cut from them. This was recognized in one of the Renaissance terms for mirror, "*sfero*" or "*spero*."[61] Globes of glass could be blown only to a certain size, and it was even more difficult to make large glass plane mirrors. Of course, a metal surface could be polished to a high luster, as it would have been in the silver sky on Brunelleschi's panel. As someone trained to work metal, Brunelleschi could have made more than one kind of mirror, and there would also have been help at hand, because in Florence (as in Bruges, as I will discuss in chapter 4), painters and mirror makers belonged to the same guilds.[62] In fact, Florentine painters had been using mirrors since the fourteenth century.[63] According to Filippo Villani, Giotto used a mirror to paint a double portrait with Dante, and, as we have seen, Dante himself was conversant in the principles of the new optics, just as Brunelleschi and his friends were intimately conversant with Dante. Early-fifteenth-century French manuscript illuminations show one of Boccaccio's famous women, the Roman Republican painter Marcia, identified by Alberti as the daughter of the encyclopedist Varro,[64] performing the delicate translation of her image in a convex mirror to a flat panel.

The Florentine architect Antonio Averlino, who called himself Filarete (Lover of Virtue), could have witnessed Brunelleschi's demonstrations as a young apprentice in the shop of Lorenzo Ghiberti, whose members were at work on the doors of the Baptistery, executing the commission Brunelleschi had failed to win in 1401. When he wrote his *Treatise on Architecture* decades later, Filarete explained perspective in terms closely paraphrasing the first book of Alberti's *On Painting*, which influenced most later treatments of its themes. Filarete's explanation centers on what he called "the plane," the trapezoidal checkerboard that is a signature of Alberti's construction—what Alberti himself called the "pavement."[65] His own debt to Alberti notwithstanding, Filarete attributed the discovery of this construction to Brunelleschi. "I believe that Pippo di Ser Brunellesco, Florentine, found the way of making this plane, and he did it according to the same reasons that are shown to you in a mirror. If you consider the matter well, you also see those mutations and diminutions with the eye."[66] That is, if we look at the reflection of a square tile in a mirror, it will be a trapezoid on the mirror's surface, and the mirror may thus serve as a guide to our own seeing. The mirror makes it clear to us just how we see what we see; at the same time, eye and mirror are again implicitly identified.

Once in possession of the "measures and reasons" of the (virtual) plane, you may, Filarete said, do as you wish—that is, you may project the form of anything. But "an even easier way to copy [*ritrarre*] everything would be to have a mirror and put in front of it those things you wish to make and look in this and you will see the contours [*dintorni*] of things more easily, and so also [you will see more easily] those things that will be nearer, and those farther away from you will seem to diminish." The mirror, in other words, transposes forms onto a two-dimensional surface and shows how such a transposition should look. *In addition*, the mirror shows us diminutions of the kind that painter's perspective makes it possible to construct. Like Manetti, Filarete associated Brunelleschi—and this construction—with the moderns; the ancients, Filarete said, may have been “subtle and acute,” but they “did not place things on the plane in these ways and with these reasons.”[67]

Filarete’s stress on the “reasons” of mirrors preserves what I think is an indispensable clue to Brunelleschi’s intent. Filarete, however, should not be understood to say that these “reasons” can be *found* by looking in mirrors but rather that they can be *verified* by looking in mirrors.[68] Although it is likely that Brunelleschi did in fact experiment with mirrors, he need not have arrived at his construction in this way. His demonstration and his construction might also have been based—and would have been most persuasive if they were based—not only on the empirical observations of reflections but also on the geometry, the “reasons” (that is, the basic catoptrics) of reflection in a plane mirror. This leads us back to the analogy of eye and mirror and back to the title of this chapter.

The subject matter of medieval optics was divided into three parts. The first is *direct vision*, when we see objects in light, which also involved the physiology of the eye itself—how is light conveyed through the structures of the eye such that we see what we see? The second part is *reflection*, when we see light *from* surfaces. The third is *refraction*, when we see forms, light from the surfaces of which has passed from one medium to another—in the standard example, from air to water, in which straight oars appear to be bent. It is often pointed out that Renaissance painter’s perspective had no direct or immediate source in medieval *perspectiva*, but it is perhaps more accurate to say that it had no straightforward source in the first part—that is, in the geometric description of direct vision. The relation to the second part of optics—namely, reflection—raises other issues and possibilities and takes us back to Alberti’s statement that some regard the eye as a living mirror, which in turn takes us back to the question of the physiology of the eye.

We have seen that the crystalline humor is diaphanous—that is, light passes through it—but also a little dense, so it is suited to the apprehension of forms in light. As Alhazen wrote, “It is proper to light that it works upon sight and proper to sight that it is passive to light.”[69] Alhazen’s translator called the interaction of light and sight *dolor*, which might mean “a pain or ache,” but as Alhazen himself explained the matter, the passion or suffering of the crystalline humor is not the kind by which the member is hurt, nor is it manifest to sense as pain or suffering.[70] Afterimages tell us that the crystalline humor is “tinctured” by sight, he said, but afterimages go away, arise only in unusual circumstances, and so finally assure us that the crystalline humor is not in fact stained. To look forward to my fourth chapter, the uniqueness of the crystalline humor’s interaction with light must remind us of the metaphors for the virginity of Mary. The crystalline humor continually undergoes the activity of light and the transmission of forms to the mind through light, but it is not changed.

We have also seen that Alhazen followed Galen in describing the crystalline as relatively flattened, either lenticular or defined by the arc of a larger sphere, as in figures 2.9 and 2.10.[71] If the seat of vision were the surface of a small sphere, concentric with the outer surface of the eye, then it would be the equivalent of a tiny convex mirror, in which case the *virtus visiva* would always give us a distorted view of the visual field, and this, Alhazen said, would be "monstrous."[72]

Alhazen was at pains throughout his argument to ensure that the crystalline humor register an image of the visual field that was true in the sense of not being inverted, reversed, or distorted. Rays therefore cannot cross inside the eye (which would create inversion and reversal) and the sensitive surface on which they fall must be regular. According to Alhazen, those rays of light entering the eye are selected that strike perpendicularly to tangents at each point on the eye's surface, while infinite others are refracted and in effect fall away as weaker than those striking directly. The more nearly planar the surface, the more nearly the order of points in the visible world on the surface of the eye is preserved and passed to the crystalline humor. If the crystalline humor is more nearly planar, in what sense is its image truer? When he repeated and glossed Alhazen's words, Witelo wrote that this relative flattening or compression made possible the better reception of forms.[73] It was better because the more nearly flat the crystalline humor, the truer its correspondence to the set of light points in the world before it.

At least from the time of Euclid, variants of the following demonstration were repeated as the first principle of catoptrics.[74] In figure 2.15 (taken from Witelo's *Optics* in Risner's *Opticae thesaurus*), *a* is a visible point above a plane mirror *ce*; *a* is reflected at point *d* to the eye at *b*, and the angle of incidence and reflection are equal relative to vertical *df*. The demonstration depends on the introduction of a "dropped" line (*cathetus*) *ac*, continued into the virtual dimension as *acg*, perpendicular to the mirror's surface, *cde*. If the line *bd* representing reflection to the eye is also extended into the virtual dimension to point *g*, then, because *ac* is perpendicular to *cde* and the two triangles share base *cd*, and because the angle of incidence equals the angle of reflection, the two triangles are congruent, and point *a* stands in the same relation to *ce*, the surface of the plane mirror, as point *g*. Or, as Witelo phrased the matter, "the point *g*, the *locus* of the image, will always be *under* the mirror at such a distance from the surface of the plane mirror as the point of the thing seen in the mirror stands above the same surface of the mirror."[75] The truth of the reflection in a plane mirror, in other words, depends on the demonstrable symmetry

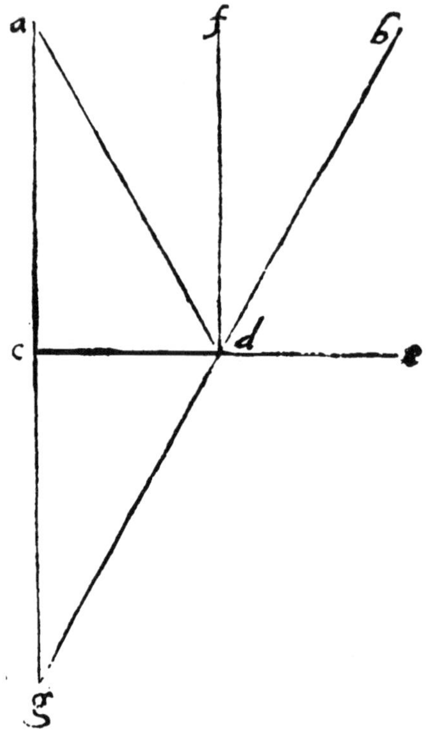

FIGURE 2.15. *Diagram illustrating the symmetry of an external point and its reflection in a plane mirror. From Witelo, in Alhazen,* Opticae thesaurus, *V.49 (p. 212).*

between a point and its image on the mirror's surface. Figure 2.16 shows the same argument applied to shape and quantity when the eye is at *e*. In fact, this principle could be infinitely expanded to apply to all points in light from the surfaces of all visible things, and everything reflected in a plane mirror might be shown to be "true" in this specific sense.

This proof of the true correspondence of the reflection in a plane mirror to any visible state of affairs thus depends on the possibility of a symmetrical construction in the virtual dimension, behind the eye if the eye is a mirror. But what exactly is the status of this verifying construction in the virtual dimension? Where does it exist? Does it exist? I will return to this crucial question at the end of this chapter. For the present, it will be enough to say that Brunelleschi's demonstration entails a simple but fundamentally important shift in the meaning of this ancient proof. The verifying construction will be acknowledged as imaginary: the argument no longer simply shows a true correspondence; rather, it validates relations *on the surface of the mirror itself*, the *apparent* quantity, which is merely a by-product of the Euclidean demonstration. The *apparent* quantity, defined by the intersection of line *ab* by the triangle *edg*, in which

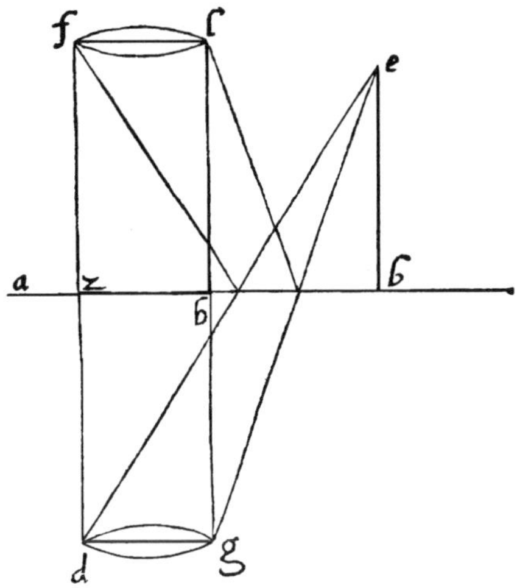

FIGURE 2.16. *Demonstration of the symmetry between shapes and magnitudes and their reflections in a plane mirror. From Alhazen,* Opticae thesaurus, *214.*

the apex *e* is the point of view, is not even labeled in figure 2.16. In fact, however, this incidental, unspecified surface and the innumerable surfaces for which it might stand will become the central reality as my argument proceeds.

A plane mirror reflection is obviously true in some ways but not in others; it may correspond point for point, as we have just seen, but *as an image* it is absolutely fugitive, dependent on any change in position relative to a viewer. It is also always small and always reversed. Writers on optics preferred to avoid such anomalies and in fact seem to have considered an argument sound when it succeeded in explaining them away. Albertus Magnus, for example, following Avicenna, noted that when we look into a convex (or plane) mirror or into someone's eye, we see ourselves reversed, right becoming left and vice versa. The optic nerve, however, which unifies the two images, "sees" the two mirrors of the eyes from the other side[76] so that the reflected world not only is in the right order but also is unreversed, as one might see the world through a window. Or problems might simply be avoided. "In a plane mirror," John Pecham wrote, "shape and quantity appear truly." By quantity, however, he cannot have meant size, which cannot appear truly. "Reflections [in a plane mirror]," Pecham continues, "retain the same order" and are true in that sense, true as *ratio*, even though left and right are reversed. This order is true even if "things in plane mirrors appear smaller than they are from the same cause as in direct vision, that is from distance."[77]

I will now turn to another participant in the conversations I have imagined, Paolo Toscanelli. A short treatise, *Della prospettiva* (On Optics) was included for many years among the works of Alberti. Alessandro Parronchi attributed it instead to Toscanelli and published it twice, the second time in facsimile.[78] Toscanelli is perhaps the most important person in the Italian Renaissance of whom we know the least.[79] He was born in 1397 and educated at Padua, returning to Florence in 1424. Toscanelli was the close friend of both Brunelleschi and Alberti as well as of Nicholas Cusanus. Manetti reported that Toscanelli knew Brunelleschi for more than forty years, which would mean that they must have known each other from boyhood. Toscanelli, we are told, considered architecture the least of Brunelleschi's excellences, merely one of the fruits of an elevated mind very different from that to be found in the run of architects.[80] In 1472—most significantly for us all—Toscanelli provided Christopher Columbus with a map by means of which one might get to the East by sailing west. Toscanelli died in 1482 and did not see this possibility realized.

Toscanelli's *Della prospettiva* is not about what we would call "perspective" at all and shows no trace of Alberti's influence, which one might expect if it had been written later in the fifteenth century. The authorities cited are all medieval—Aristotle, Alhazen, and Giovanni Sacrobosco[81]—and the text explains optics according to the traditional division I have mentioned, beginning with direct vision, followed by reflection and refraction. The treatise is addressed to someone called Polixeo, who wants to know something about optics; it is clear and instructional in tone, with many concrete examples; and, especially when compared to the great treatises from which it derives, its arguments are simple and introductory.

As historians with other interests have noted, clouds began to appear in the skies of Florentine painting at just about the time perspective became an issue. Giotto's lapis lazuli skies give way to the haze and weather on the horizon of Masaccio's *Tribute Money* (figure 2.17) and, more remarkably, in sculpture, in Donatello's *St. George* relief (figure 2.7). At a certain point, Toscanelli asked how a large object, such as a cloud, could be truly seen in a small mirror. We may fix this question in our imagination with a vivid example from Lucretius: "A puddle of water no more than one finger deep, lying between the stones upon a paved street, offers a view downwards under the earth to as great a reach as the open heavens yawn on high, so that you seem to look down upon the clouds and heaven, and you see manifest objects miraculously buried beneath the earth."[82] Since he said the height of sky and apparent depth of reflection are equal, Lucretius must have known his catoptrics, and Toscanelli also offered a version of the ancient argument we have considered proving the truth of reflections in plane mirrors. In Toscanelli's argument, a *quantity* is again shown to be symmetrical in the virtual dimension, and (as in figure 2.16) two *catheti* are necessary to define that quantity. His argument goes as follows. If we lay a

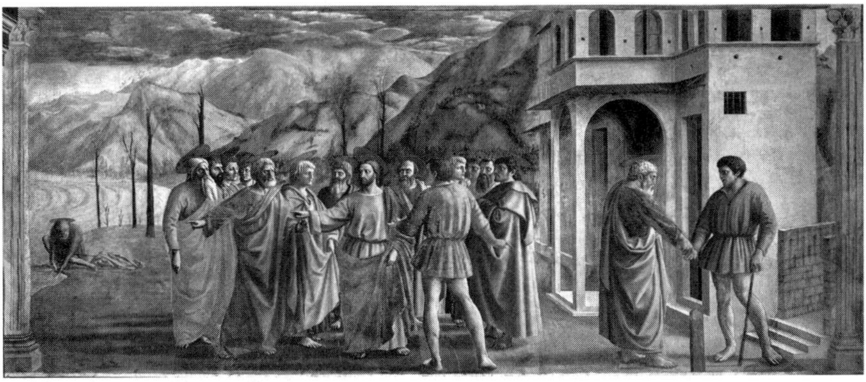

FIGURE 2.17. *Masaccio,* Tribute Money, *fresco, 1420s, Santa Maria del Carmine, Brancacci Chapel, Florence. Scala/Art Resource, New York. (Reproduced in color as plate 6.)*

plane mirror face up on the ground, we may extend the plane of the mirror over the ground to become what he calls the mirror's "equality." If *fl* in figure 2.16 is a cloud, and *ab* is the plane of the mirror on the ground and its "equality," then *catheti* representing the quantity of the cloud *fl* may be dropped (*fz and lh*) to this plane, and everything once again falls into place in relation to the eye at *e*. When the diagonal lines defined by the angles of reflection from *f* and *l* to the point of sight *e* are extended from *e* into the virtual dimension, they in effect become equivalent to the original lines of incidence, striking the *catheti* at points *d* and *g*, equivalent to the points *f* and *l* defining the breadth of the cloud. Relative to the eye (*e*), the triangles formed are once again congruent and symmetrical with those defined by the two *catheti* and the angles of incidence and reflection.[83] Similarly, the eye—or more precisely the crystalline humor—might be Toscanelli's small plane mirror on the ground or the puddle of Lucretius, and in either case we might prove that what we see is symmetrical with reality.

For Toscanelli, it would have been significant that the same argument holds whether the *catheti* fall on a mirror's surface or on its "equality," and the mirror need only be large enough to reflect the points of incidence and reflection on *ab*, which define the size of the cloud *as it appears*. Any point, all points adding up to whatever might be seen are in fact subject to this argument. We may easily imagine a recessive plane with multiple clouds, scattered clouds, like the miraculous snow-bearing pillow-clouds of Masolino's *Founding of Santa Maria Maggiore* (figure 2.18).

I remarked earlier that the *apparent* measures and the ratios on the reflective surface of the mirror, rather than simply being by-products of the demonstration of the symmetry of world and reflection, have assumed a new importance—or, better, a new reality. "Is it possible," Toscanelli asked, "to find the point of reflection in any mirror?" "Many," he said, argue that this is impossible because "a [point] is an invisible [that is, a purely geometric] thing." But Toscanelli met this objection by insisting that the point of reflection is not a geometric point but is instead a *physical, material* point that may be represented by other physical and material points. "I say that the place of reflection, of anything . . . may be found most easily." Making sure that eye, mirror, and object remain stationary, "take a tiny bit of wax or paper," and stick it to the surface; this will mark "the place of the incidence or reflection upon the mirror; do similarly with every other part, and in any place that may be."[84] This reply is a nominalist reply: one material point may simply be marked by another, and a set of such points on a mirror's surface would be a true reflection of the world before the mirror.

We may now return to the question of the status of the symmetrical, verifying reality "within" the mirror. This reality, Toscanelli continued, does not multiply rays—we could not reverse these diagrams, exchanging "real" and virtual space—and therefore it is not a true body; it is not itself seen. But "although within the mirror the space is not true, this does not matter, because the similitudinary things do not occupy space. . . . And you have manifest examples of this when in dreams you see various fantasies that seem true bodies, in a true, luminous place, and nonetheless it is not so."[85] By this, Toscanelli did not mean that the demonstrable image in a mirror is simply illusory; rather, he meant that it belongs to another order of existence altogether, to that which can be imagined and manipulated in imagination, and in these terms may be "thought about." Such a position might seem unexceptionable now, but it was controversially progressive at the time. In the vituperative poem against Brunelleschi mentioned earlier in this chapter, his erstwhile friend Giovanni da Prato took the antinominalist, realist position, accusing Brunelleschi of making "uncertain things certain" and of "fantasy."[86] This is an inversion of the anti-Platonic Vitruvian characterization of *skenographia* considered in the last chapter.

We may now briefly and finally turn to the question of why Brunelleschi treated construction and reflection as equivalent in his panel. He must have meant his constructed buildings to be true in the same way that the reflected sky was true and so put both on the same surface to be united in a single planar reflection. Although it is clear from Manetti's description *how* Brunelleschi's first panel was to be used, it is not clear *where* it was to be used. Was the viewer meant to stand where Brunelleschi had stood, to look through the panel at the handheld mirror, then take the mirror away to see the same buildings and sky? Or did Manetti mean that wherever the panel was looked at under the

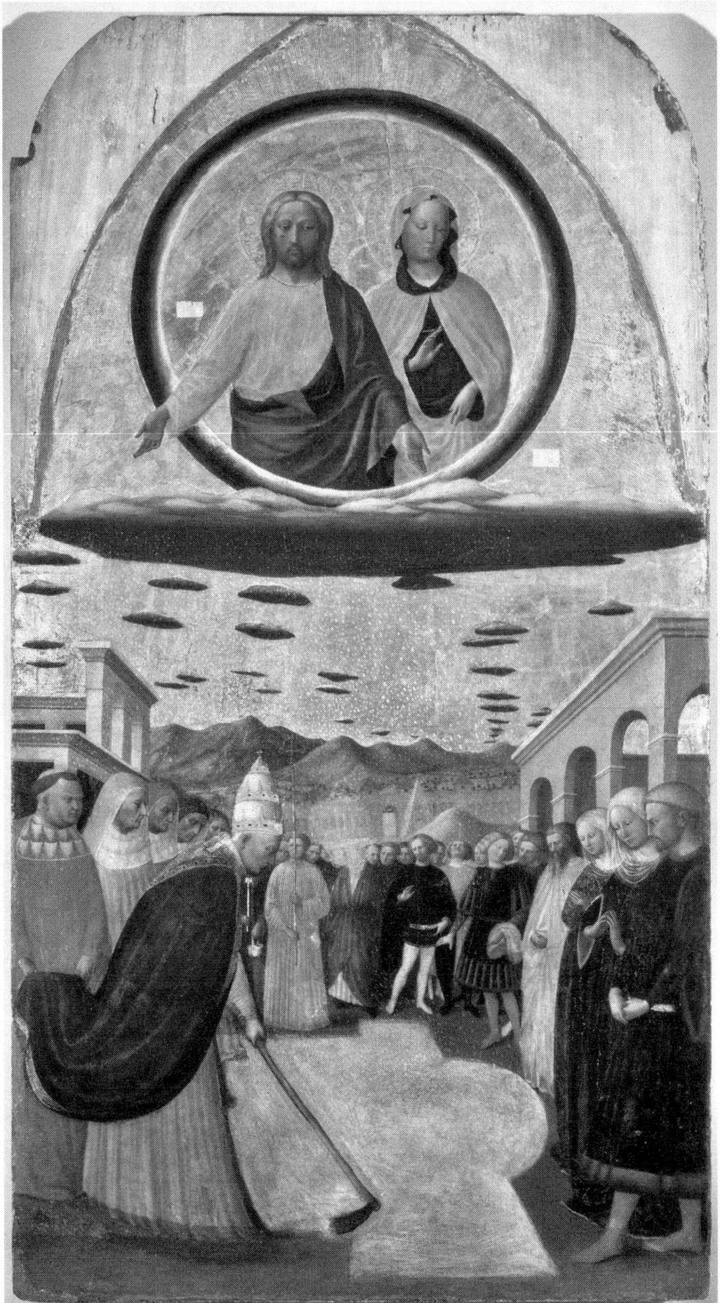

FIGURE 2.18. *Masolino,* Founding of Santa Maria Maggiore, *wood panel, ca. 1425, Museo Nazionale di Capodimonte, Naples. Scala/Art Resource, New York.*

conditions Brunelleschi had set, it would have been possible to see the deeply familiar east side of the Baptistery *as if* from the central portal of the Duomo with whatever sky happened to be above whatever place the viewer happened to stand? The second possibility seems more likely. From the actual standpoint, the contours of the buildings according to the construction I have described could not coincide with the reflected sky, and the illusion could thus not have been seamless. It is significant that precisely this objection was answered in Brunelleschi's second panel, showing the Palazzo della Signoria from the opposite corner of its piazza. The second panel was too large to be held at arm's length. The construction of the buildings was presumably the same, using a measured plan and elevation, but this time the sky was cut away altogether so that the truth of the projection of the buildings was demonstrated by the coincidence of the painted profiles against the real sky. The construction could remain the same, but the mirrors were unnecessary. The true reflections in the living mirror might extend to the whole world.

At this point we may return to Filarete, who followed up his explanation of perspective with a defense. "You may say this is false which shows you something that is not [*una cosa che non è*]. It is nevertheless true in *disegno*, and true because *disegno* itself is not a true thing, but rather a demonstration [*dimostratione*] of that thing you copy, either that you copy, or that you wish to demonstrate. So then this is true and perfect according to *disegno* and without it one is not able to practice the arts of painting and sculpture well."[87] What is it that is not? And what is it that is demonstrated? It is the *apparent* image and more specifically the image on the surface of the mirror, the image that, as we will see in Chapter 4, is embraced by Narcissus, the inventor of painting. It is the peculiar reality of that image that is demonstrated by perspective and by art, by *disegno*—that is, both by vision and by imagination.

In general, Euclid's geometry of vision, perhaps formulated under the pressure of the ideal of geometric purity, softened over the centuries, yielding to physiology and the study of light. Alhazen and writers after him considered the visual angle a convenience, a notional approximation. The visual pyramid, Pecham wrote, is a truncated pyramid that does not really come to a point in the eye; the point exists only in imagination.[88] This convergence is prevented by the living surface of the eye. The "point" of view is thus notional, mental, mathematical, and vision truly takes place on the sensitive surface of the eye. By the same token, if the crystalline humor is relatively flattened, it is not a plane. Vision is not notional; the seat of vision has the particularity of all flesh, and to recast it as planar is to identify it with notional vision, with the putative vision of geometric demonstration. It is only when we see the world before us as if in a plane mirror, from a point which is as if it were itself within the notional world of thinkable and imaginable metric relations, that it comes into view with the notional purity of the forms of Brunelleschi's own architecture (figure 2.8) or of Piero della Francesca's imagined architecture of sacred history.

CHAPTER 3 *Speculum Mundi*

In 1435, Philip the Good of Burgundy upbraided his treasurer for refusing to more than triple Jan van Eyck's salary, writing that the painter's "art and science" uniquely suited him for "great works."[1] If this is unquestionably a record of the duke's high esteem for van Eyck, it is a terse one from which it would be unwise to try to extract too much: the terms "art" and "science" have always had slippery connotations and could be used more or less precisely then, just as they are now. And it can only be guessed what kind of "great works" the duke might have had in mind, since his painter also had important responsibilities other than painting. I will, however, continue the general argument of these chapters by suggesting that the duke meant to refer not just to van Eyck's mastery of painting but also to his fluency in the new overlapping theories and practices of optics and its related disciplines. This fluency leads to another of my themes: that the middle sciences, what Leonardo called the geometries of visual lines and what I have called a common specialist culture—everything from surveying to cosmology—constituted a kind of lingua franca in terms of which the great accomplishments of Italian and Northern Renaissance art were local but also mutually intelligible. The question is thus not who had "true" perspective and who did not or did not quite—the question that stubbornly haunts modern scholars[2]—but rather who made what choices and followed up what possibilities in this ongoing specialized culture. And so, when the humanist Bartolommeo Fazio praised van Eyck as "not unlettered, particularly in geometry and such arts as contribute to the enrichment of painting," he should perhaps be understood to mean that van Eyck was also knowledgeable and skilled in a number of more practical geometric disciplines related to the art of painting.[3]

The history of art still bears the strong imprint of nineteenth-century nationalism, and the search for national "spirits" still tends to conceal common European culture. Accordingly, the Italian Renaissance attitude toward northern painting continues to be exemplified by remarks attributed to Michelangelo by the young Portuguese Francisco de Hollanda, remarks that, if Michelangelo actually made them, would have been made around 1538.[4] Michelangelo, then in his early sixties, dismissed northern painting as mawkish and superficial, without the substance of *disegno*, appealing only to the outer eye. Perhaps Michelangelo remembered with a twinge of adolescent jealousy the appreciation of his Medici patrons for Flemish painting.[5] However that may be, Michelangelo's attitude was clearly the exception rather than the rule, coming as it did at the end of a long history of admiration for Flemish painting in Italy. And, as we shall see at several points in this and the next chapter, admiration was firmly rooted in vigorous, well-established economic and commercial relations.

As an example, we may consider Jan van Eyck's famous *Arnolfini Portrait* (figure 3.1). Its central convex mirror aside, the marriage of an Arnolfini to Giovanna Cenami in 1434 would have merged two powerful Lucchese families, and if this wily-looking man is Jan Arnolfini, he had the strongest connections to Philip the Good, serving as his counselor and financial adviser on the way to becoming the governor of finance of Normandy.[6] Jan van Eyck had other Italian connections, and if his portrait of a musician (if it is a musician and not just a God-fearing man) labeled *Tymotheus* is Guillaume Dufay, as some have suggested, this Burgundian composer spent much of his life in Italy and wrote the music for the dedication of Florence Cathedral by Eugenius IV in 1436.[7] The name "Timotheus" recalls one of the great musicians of antiquity, but Dufay was associated with an *ars nova* of music. Alberti also looked back to antiquity, but in the Latin version of his treatise *On Painting*, finished in 1435, he wrote of the art he described as unprecedented, set out in entirely new terms, *novissima*, invented by Narcissus.[8]

Michael Baxandall called Fazio's *De viris illustribus* (from which I quoted praise for Jan van Eyck a few paragraphs ago) the culmination of the humanist criticism of art.[9] Fazio was born in 1405 in La Spezia, on the coast southeast of Genoa, where Alberti had been born the previous year. Fazio was a pupil of Guarino Guarini and, like Alberti, moved in high scholarly, ecclesiastical, and political circles. After 1444, Fazio was in Naples, in the two-year-old court of Alfonso V of Aragon (Alfonso I of Naples), whom Fazio served as historian and secretary. Already in the fifteenth century, Fazio was identified with the title character Momus, the god of mockery in Alberti's *Momus*, a fantastic and mordant satire written around 1450.[10] In his *De viris illustribus*, written between 1455 and 1457, the year of his death, Fazio listed Alberti among his illustrious contemporaries, specifically mentioning his treatise *On Painting*. Perhaps persuaded by Alberti's arguments for the dignity of the art, Fazio included sections on painting and on four painters—two Italian, Gentile da Fabriano and Pisanello, and two northern, Rogier van der Weyden and Jan van Eyck. Fazio extolled van Eyck as "the prince of painters of our time."[11]

When Fazio described Jan van Eyck as "not unlettered," this might be understood to be a left-handed way of saying van Eyck *was* in fact learned but not, like Fazio himself, a humanist. Fazio wrote not only that van Eyck was learned

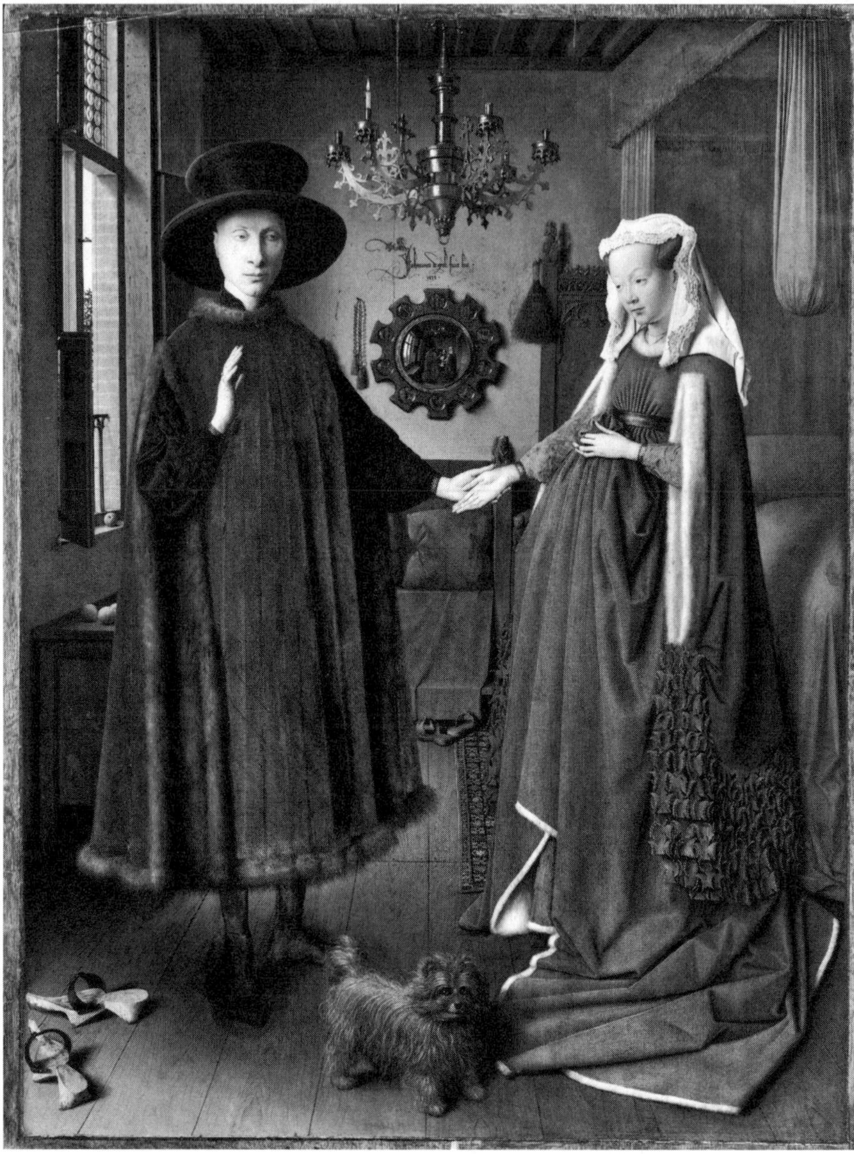

FIGURE 3.1. *Jan van Eyck,* The Arnolfini Portrait, *oil on wood panel, 1434, The National Gallery, London. © The National Gallery, London.*

in geometry and arts related to painting but also that his learning extended to the many things he had discovered "about the properties of colors recorded by the ancients and learned by him from reading Pliny the Elder and other authors." This may be compared to Alberti's stipulation that the painter is to be learned in the liberal arts "and above all to have a good knowledge of geometry."[12] Like Fazio, Alberti immediately referred to Pliny the Elder but more specifically to Pliny's account of the learned Pamphilus and his equally learned pupil, Apelles.[13] The geometry necessary to understand the "rudiments" of his new art, Alberti said, was not difficult but was indispensable. If this chain of associations might ultimately compare the best modern painter to Apelles, it should not be inferred that just because a tireless *topos* is at work, both Apelles and van Eyck were not really learned in geometry (as we know Alberti to have been) or that the "geometry" in question was not also optics and in general part of the common specialist culture for which I have been arguing.

It is likely that Fazio's choice of painters and paintings reflected the taste of the Neapolitan court and its prince, Alfonso I, although the taste of court and ruler must also have been influenced by Jan van Eyck's broader reputation. Partly because of his own extensive travels, the influence of Jan van Eyck and his painting was strong in Spain, and Alfonso collected van Eyck's paintings, sometimes with Fazio as an intermediary, as in the case of the first painting Fazio described, a triptych painted for the Genoese merchant Battista Lomellini. This triptych, like the other two paintings by van Eyck Fazio described, is lost, but Fazio claimed to have known all three well. He ended his long paragraph with the statement that van Eyck "is said to have made many other works, about which I have not been able to obtain full knowledge,"[14] suggesting that he was willing to write only about paintings of which he had firsthand knowledge.

I will examine other remarks by Fazio on van Eyck's paintings in the next chapter. For the present, I will concentrate on the second work he described, a map of the world, and part of a third, a landscape. After describing the Lomellini triptych, which was in Alfonso's private quarters, Fazio continued as follows: "Also his is a map [*comprehensio*] of the world in circular form" made for Philip the Good. Fazio reserved special praise for this map, writing that "it is thought that no work has been done more perfectly in our time." It was, in short, the finest work of the leading painter of the age. Maps themselves have only recently begun to receive art-historical attention,[15] and lost maps defy easy construction in any case. It has even been urged that, Fazio's enthusiasm notwithstanding, the map should be dropped entirely from the list of van Eyck's lost works.[16] The astronomer Guillaume Hobit was paid in 1444 for a "world-map after Ptolemy on which he worked for three and a half years," in which case he and van Eyck, who died in 1441, could only have collaborated for a brief time.[17] Even if that is so, however, the Duke of Burgundy might have had other world maps painted, and Fazio should be taken seriously because he was well informed regarding an extraordinary painted map of the world to which Jan van Eyck had made a substantial and evident contribution.

The map Fazio praised so highly might have been painted—and others might have been made as well—at various times during the sixteen years van Eyck spent in the Burgundian court between 1425 and his death.[18] *Mappamundi* (maps of the world) were associated with rule at least from the time of Augustus and were prized possessions of Renaissance rulers—the duc de Berry had several,[19] as did Cosimo de' Medici[20]—and, as we are about to see, maps were of more than usual interest to the Dukes of Burgundy.

The disappearance of van Eyck's map is perhaps to be explained not simply by the fragility of works of art and the accidents of history but by the rapid obsolescence of world maps at the beginning of the age of European expansion. In the 1420s, the Portuguese were sailing down the west coast of Africa, beginning the transformation of the millennial Mediterranean view of the world, and cartography was changing rapidly. We might consider the fate of the Lorenzetti *mappamundi* in the Palazzo Pubblico in Siena, which was greatly admired in the Renaissance, although, as Ghiberti already observed, it could only have been less than perfect because the *Geography* of Ptolemy was not known when the map was painted. It had fallen to pieces by the eighteenth century.[21]

I have suggested and will press the point that when he made his lost map, Jan van Eyck was conversant—or could easily have become conversant—in then current principles, practices, and issues of the new cartography, and I will finally make the historical critical argument that this cartographic experience and understanding, this awareness of the possibility of the representation of virtual space of an entirely new clarity and scale, stands behind van Eyck's extraordinary landscapes, which are so fundamental to that modern genre and beyond that so fundamental to pictorial meaning and expression as they have come to be understood.

As I shall discuss in detail, Fazio specifically noted that in van Eyck's world map, distances between places could be determined by measurement, and in the next painting Fazio described, showing "women of uncommon beauty stepping from the bath" (which I will discuss briefly in the next chapter), there was a window through which were to be seen "horses, minute figures of men, mountains, groves, hamlets and castles, carried out with such skill that you would believe one was fifty miles from the other."[22] That is, through this window was to be seen a landscape in which far distances seemed measurable, much as they were in the great map. Jan van Eyck had painted—as indeed he *did* paint—landscapes of a new kind and, more specifically, landscapes of cartographic scale, precision, and credibility.

Jan van Eyck was about thirty-five years old when he joined the court of Philip the Good in May 1425. The duke himself was about thirty.[23] In September, the second duchess of Burgundy, Bonne d'Artois, died after a short marriage, and three years later, in 1428 and 1429, van Eyck was a ranking member of a Burgundian delegation to Spain and Portugal, where, among other activities, he painted Isabella, the infanta of Portugal, sister of Prince Henry the Navigator. The wedding of Philip and Isabella took place on January 7, 1430, finally to assure the duke's succession. In the years between 1425 and 1428, however, van Eyck was also employed in other ways. By July 14, 1426, he had participated in a pilgrimage on the duke's behalf, and by August he was on the road again, off on a "certain distant secret voyage," for which he was reimbursed at the end of October. Some agreement seems to have been reached that van Eyck's pilgrimage is most likely to have taken him to the Holy Land.[24]

What might it have meant to make a pilgrimage to the Holy Land on the duke's behalf? People went on pilgrimages for various reasons—out of piety, fear of Purgatory, or simple penance, the last of which might be self-generated or imposed. Tourism and adventure were factors, and aristocratic entourages were often opportunities for ostentation and sumptuous display. Although he was a major donor to Christian shrines in Jerusalem, and traffic between his court and Jerusalem was fairly routine, Philip never visited the holy places himself, and Lotte Brand Philip has suggested that van Eyck, like other members of the court, went there as a kind of proxy as well as to record the sights for the duke. More specifically, she suggests van Eyck might have gone in the delegation sent to Jerusalem in 1426 under Guiot, the Bastard of Burgundy.[25] Eyewitness of the holy places and records of eyewitness would certainly have been valued, but such exact observation might also have involved reconnaissance, part of which is mapping.

Cartography may be disinterested, but mapping may also be preparation for military action and occupation. There are long traditions of mapping for legal disputes involving land boundaries, and maps facilitate trade. But maps are and long have been of fundamental strategic importance: we still closely associate cartography with "gathering intelligence." Against this background, we may understand another journey sponsored by Philip the Good and cosponsored by Henry V of England in 1421–23. During these years, another member of the Burgundian court, Guillebert de Lannoy, traveled to Prussia, Poland, Ukraine, the Crimea, Constantinople, Rhodes, Jerusalem, Cairo, and Crete. This trip was clearly more than pious tourism. Crete and Rhodes were important staging areas for Crusades, and the Armenian Christians as well as their sometime allies the Mongols figured prominently in European military planning. Egypt was considered by many to be a crucial first step toward the Holy Land, as may be seen in the defeat of the army of Louis IX of France in the Nile Delta in the spring of 1250. Jan van Eyck's pilgrimage thus belongs to the series of such journeys by members of the Burgundian court in the 1420s, and he was sent on another "secret distant" mission in 1436. To return to our theme of cartography, when Jan van Eyck joined the Burgundian court in 1425, he might be supposed to have taken his place in an ongoing cartographic project, using information to be verified and expanded while on his own "pilgrimages." And this would suggest that van Eyck actually practiced cartography as the basis for the more symbolic and theoretical world map of which I will try to provide some imaginary glimpse in this chapter.

As the Ottoman Turks consolidated their conquests on the eastern frontier of Europe through the fourteenth century, they were resisted by, among others, the Dukes of Burgundy, beginning with Philip the Bold, grandfather of Philip the Good. Victory in a Crusade would have meant unrivaled prestige as the dukes sought higher, royal status among the rulers of Western Europe, and Philip the Bold played a central role in the Crusade of 1396, which, however, ended in military disaster when the Christian armies were defeated by the Turks at Nikopolis. Worse yet, Philip the Good's father, Jean de Nevers, who was to become John the Fearless, was captured in this devastating defeat and ransomed for an appropriately princely sum. Philip the Good was thus born into a family tradition of crusading[26] and in addition to whatever religious motivations he might have had sought status, revenge, and perhaps restitution of the enormous price paid for his father. Philip's interests were reinforced by his wife, Isabella of Portugal (her brother, Prince Henry the Navigator, was another zealous advocate of Crusade), and the Burgundian court favored chivalric literature and schemes for new Crusades. At the popular level, the Holy Sepulchre in the hands of nonbelievers could always be compared to the humiliation and suffering of Christ, thus to stir support for Crusades.

Philip the Good is remembered as a great patron, especially of illuminated books, but his fascination and that of his ostentatious court with chivalric themes was also an insistence on the theme of the Christian ruler as crusader. All his planning, however, culminating three generations of rule, only led to another military disaster, the defeat of the Christian armies at Varna on the Black Sea in 1444.

But to return to Jan van Eyck's map, Fazio wrote that in it "you may distinguish . . . not only places and the lie of continents, but also, by measurement, the distances between places."[27] What might he have meant when he said you may distinguish by measure (*dignoscas metiendi*)? He certainly meant to say that at least part of the map was made to scale, in which case it must have been based on a grid, perhaps a system of longitude and latitude. As we are about to see, this would have placed van Eyck's map at the forefront of cartography and would at the same time also have placed it in longer cartographic traditions. Maps played a persistent role in the history of the late Crusades, and, although the Crusades were conducted over relatively familiar ground, the *oikumene* (the known, habitable world), the precise intelligence maps might provide was not available, and the means for such planning were only taking shape. So, if the later Crusades were failures and were not themselves voyages of discovery, they were at least a stimulus to the development of the cartography that would guide European expansion westward rather than eastward or to the East by sailing to the west.

Twenty years after the Church sanctioned its mission to the East at the Council of Lyon in 1245, Pope Clement IV, concerned about the technological inferiority of the Latins, asked the Franciscan polymath and protoscientist Roger Bacon to send his writings to Rome at once. As part of the great project of setting down his ideas, Bacon prepared a map of the world, now lost, based on longitude and latitude. He described a projection—here reconstructed by David Woodward (figure 3.2)—in which meridians marked along the equator converge on the North Pole in a manner roughly similar to the projection of Ptolemy I will discuss shortly.[28] Although Bacon did not have Ptolemy's *Geography*, global geography and astronomy entail one another, and his scheme must have been based on Ptolemy's *Almagest* and its Arabic tables and commentaries. For Bacon, of course, astronomy included astrology, and the precise charting of the world, he wrote, was necessary in order to understand and convert the peoples to be encountered in their own star- and planet-influenced conditions, places, and times. His map, in short, putting the surfaces of the habitable world in specific relation to the powers of the heavens, would have provided the Church with a kind of astrological global positioning system. Bacon also noted an important obstacle, however: few positions had been established from which reliable maps could be made.

The fall and destruction of Acre in 1291, which marked the end of Latin presence in the Near East, also saw the beginning of a new literary genre, books on how to recover the Holy Land. The first of these, entitled *Liber recuperationis terrae sanctae*, was written by the Franciscan Fidenzio of Padua and dedicated to

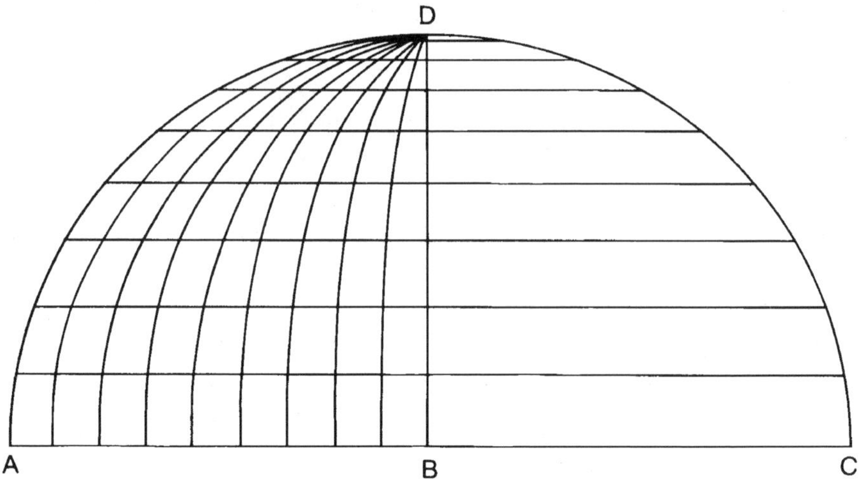

FIGURE 3.2. *Reconstruction of the projection used in the map appended to Roger Bacon's* Opus maius *(1268).* ABC *is the equator, and* BD *is the prime meridian. From Harley and Woodward,* History of Cartography, *1:322, fig. 18.28. © University of Chicago Press.*

Pope Nicholas IV, who was shortly to solicit and receive other proposals. These books offered histories, strategies, and logistics as well as evaluations of routes and resources.[29] The most elaborate, by the Venetian Marino Sanudo, was the *Liber secretorum fidelium Crucis*, presented to Pope John XXII at Avignon on September 24, 1321. All copies of this book contain a map of the world (figure 3.3), together with more detailed maps of crucial regions. Sanudo was assisted in his project by a cartographer, Pietro Vesconte, who figures importantly in the broader history of mapmaking.[30] Vesconte also made the map of the Holy Land in figure 3.4 for Marino Sanudo around 1320. Both maps are oriented in the traditional European Christian manner I shall discuss, with east at the top. The coast of the Mediterranean in figure 3.4, labeled the River Jordan by a pious, much later hand, is shown as a north-south line—that is, from our left to right. We are told in the legend that the squares of the grid stand for measures of one league or two miles, so that distances between one place and another are easily estimated. There are significant landmarks, mountains—many at the top, at the horizon, where the grid ends—and there are also bodies of water and place-names. To the degree that this is an accurate map, it must be supposed that the distances represented were either paced off or otherwise determined, perhaps surveyed, at the various sites.

Vesconte was one of the first to make grid maps, and he was also a maker of portolan maps. The origin of portolan maps is mysterious and disputed, but

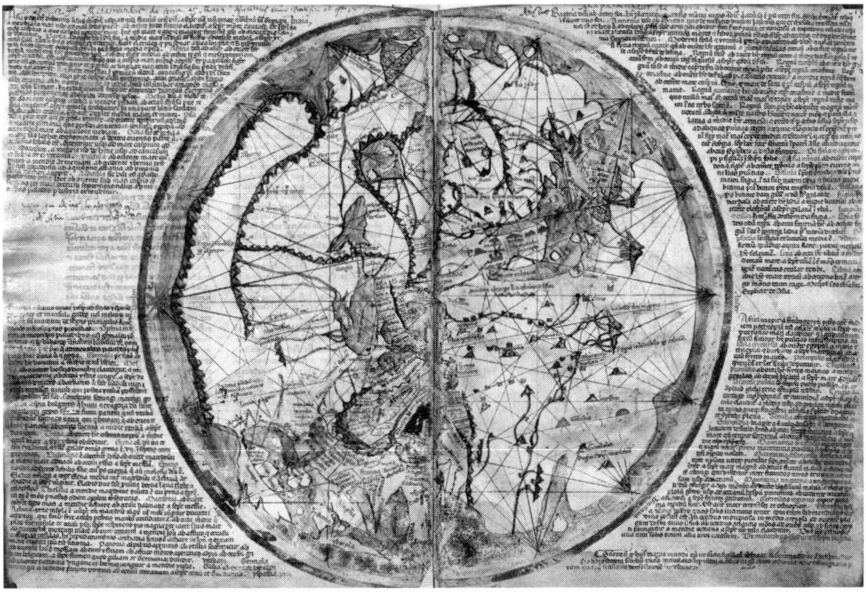

FIGURE 3.3. *Pietro Vesconte,* World Map, *ca. 1320, British Library, London.*

they were functional and descriptive, providing newly precise guides to shorelines of the Mediterranean and Black Seas. They were, in other words, used for navigation and trade in the world known to Europeans, and they provided the means by which that world could be gradually expanded and effectively addressed. Although they were accurate to a degree especially remarkable in comparison to earlier medieval maps, portolan maps were only occasionally drawn to scale. Still, in general, European cartography progressed through the fourteenth and fifteenth centuries by successive additions and corrections provided by portolan maps. In his world maps for Sanudo, Vesconte extended the signature system of rhumb lines and wind roses over land, perhaps to convert commercial to expressly military purposes.

Although these more practical portolan maps could be very accurate and could be pieced together to show larger areas, they were not world maps, and a more general principle of organization was necessary in order to coordinate the representation of all the world's parts. This brings us to the second-century *Geographia* of Ptolemy. Like the thirteenth-century map—or framework for a map—of Roger Bacon (see figure 3.2), Ptolemy's map was based on a network of latitude and longitude relative to which places could be located. This framework, aligned to the north, is still what we expect of a map. Figure 3.5 is from a thirteenth-century Byzantine Greek manuscript of Ptolemy's world map, made

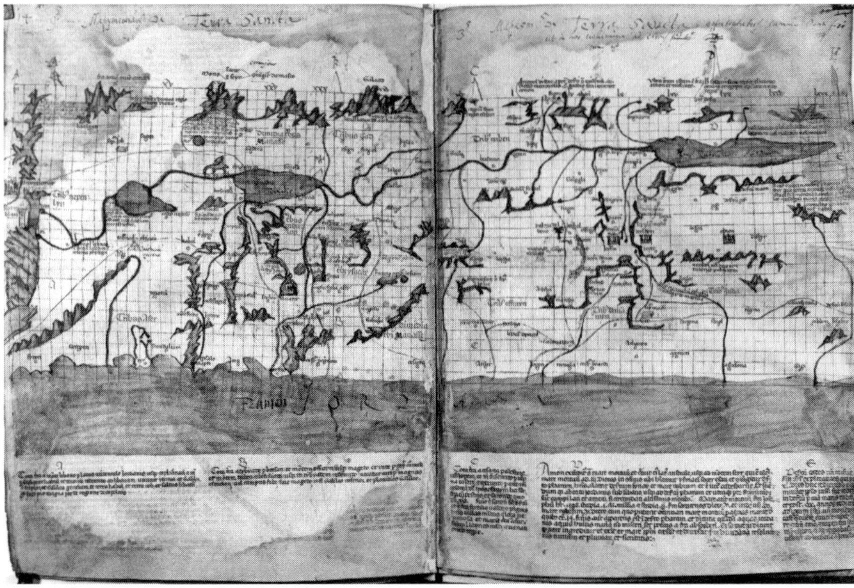

FIGURE 3.4. *Pietro Vesconte,* Map of the Holy Land, *ca. 1320, British Library, London.*

according to his first projection, with straight longitudinal lines. Such maps often were not drawn from earlier maps but from Ptolemy's list of places in relation to coordinates.

Ptolemy's *Geographia* was brought to Florence around 1400 and translated into Latin by 1410, after which its copying and illustration became a flourishing industry in that city.[31] But Constantinople was for all intents and purposes under siege through the first half of the fifteenth century, and the Byzantine Empire had a vital stake in facilitating Western crusades (the calamitous Fourth Crusade notwithstanding). Thus, if the Crusades yielded few absolute advances in geographical knowledge, they may still have contributed to the important result of awakening interest in Ptolemy and cartography, which quickly spread throughout Europe.

The first map of a European city drawn to scale—existing in a later fifteenth-century copy—was made of Vienna in 1422. In fact, Vienna was a major center for the study of both cartography and closely related astronomy. For two decades beginning around 1420, such work was directed by the astronomer and mathematician Johannes de Gmunden at the University of Vienna and by Georg Muestinger at the Augustinian monastery of Klosterneuburg. World maps were made together with the sky maps necessary to plot locations, and experiments were made with terrestrial globes (which can in any case only have shown the traditional world of three continents). The first sky maps were made around 1425 by Conrad von Dyffenbach. What survives from the earliest phase of this effort are tables—apparently made without knowledge of Ptolemy's *Geographia*—that plot the locations of places in Europe and Africa according to a metric coordinate grid of latitude and longitude.[32] By 1430, instructions also existed with the title "*Instrumentum* for the Distances of Cities and Kingdoms, called Cosmographia." With this instrument it was possible for princes, "for their use, honor and good reputation" and "for the decoration of their palaces and churches," to make "the *pavimentum* here described, in which may be seen the quantity of the entire inhabitable world, with every climate and right way of traveling from one region or city to another; the distances of every region and city [may be measured] easily and without any tedium of mind."[33] We are getting closer to Fazio's description of Jan van Eyck's map.

Between 1425 and 1430, this Viennese group incorporated its new measurements into a more ambitious and comprehensive world map adapting the traditional Jerusalem-centered OT scheme (which I will explain shortly) with Ptolemaic and portolan maps. The map made by the Benedictine Andreas Walsperger in Constance in 1448 (figure 3.6) is one of the echoes of this project.[34] In this map south, following the Arabic (and Chinese) pattern, is at the top; east is then to our left, marked by a great cathedral-like Paradise, the source of four rivers. Christian cities are marked in red, others in black. The map's legend reads as follows: "Wishing therefore to know in the present figure how many miles one place or city may be situated from another, take a compass and place its foot at the middle point with the name of any city marked . . . and extend the other foot to any other city you may wish. Then place the compasses so extended on the scale [*scalam latimetram*] here below." On this scale, Walsperger tells us, the measures are ten Teutonic miles, each mile is ten thousand paces, and each pace is two feet.

This map is a cosmos, its successive rings passing from earth and water to the quintessence. The disk of the habitable world, although clearly compass drawn, is also off center. Ocean is broader to the north, and, as if he had heard of the Portuguese voyages around the west coast of Africa or seen Arab maps of southern Africa, Walsperger extended Asia and Africa into the first rings. This slight eccentricity itself suggests that the imaginative expansion from habitable disk to the unknown terrestrial globe was under way. In any case, the instructions for determining distances on this map take us back again to Fazio's description of Jan van Eyck's map, which might have had a similar ancestry.

Walsperger's map is what historians of cartography call a "transitional map" of the world, and so, I will argue, was Jan van Eyck's. In order to understand

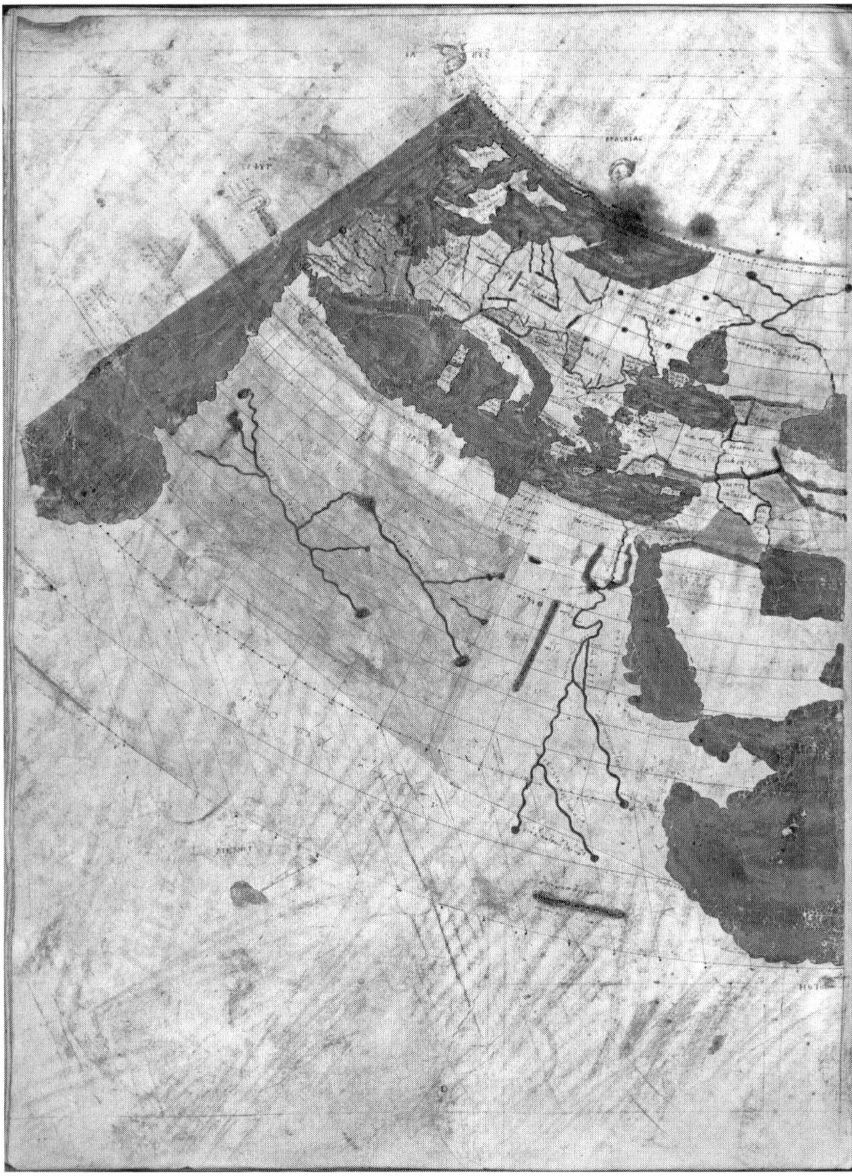

FIGURE 3.5. Map of the Known World According to the First Projection of Claudius Ptolemy, *Byzantine, thirteenth century, Biblioteca Apostolica Vaticana, Vatican, Rome.*

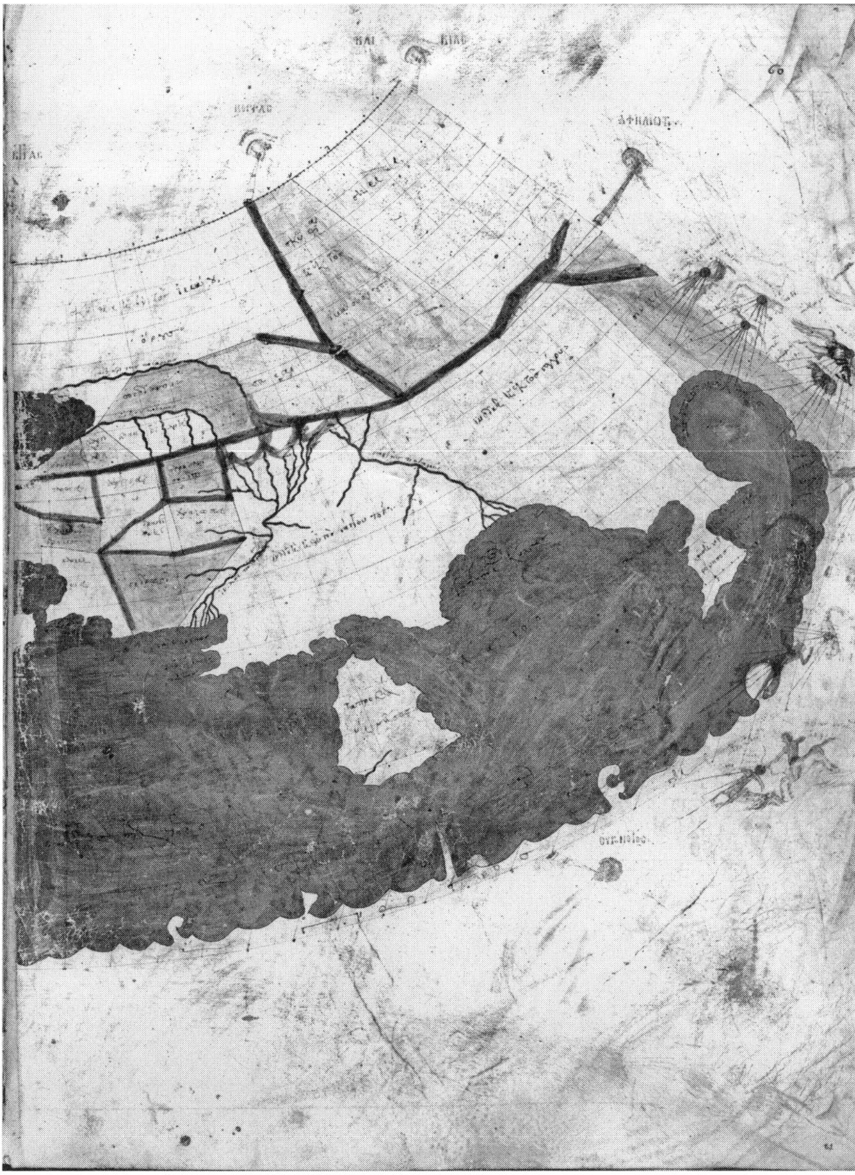

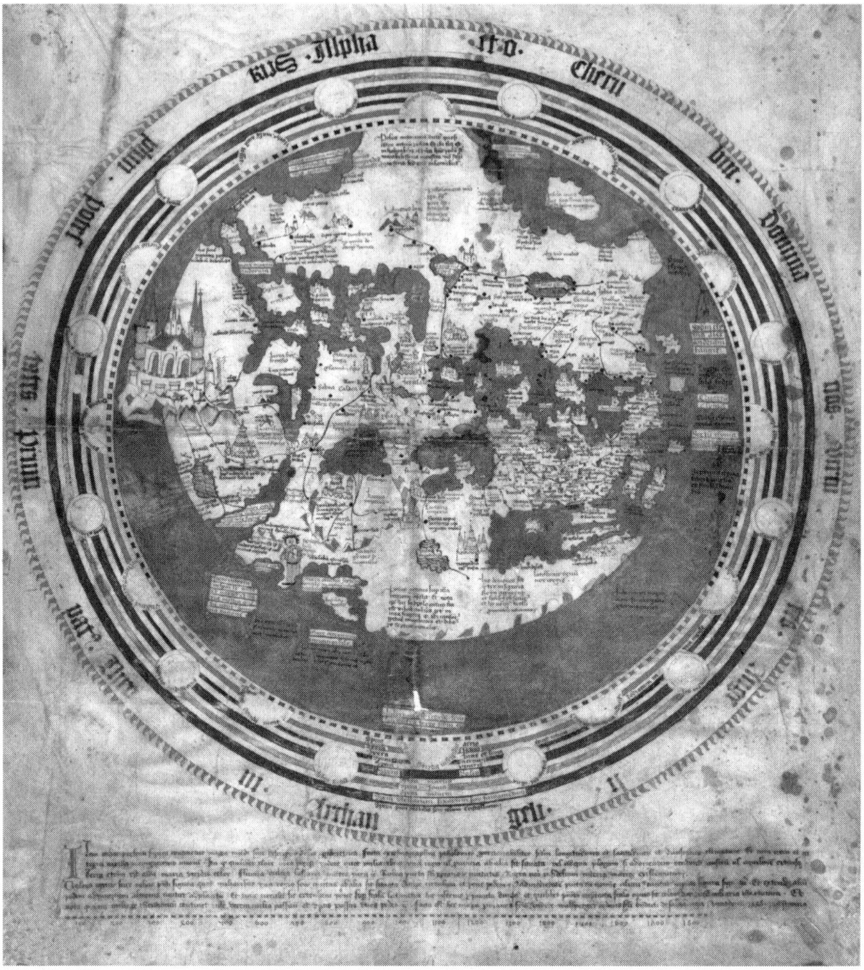

FIGURE 3.6. *Andreas Walsperger,* World Map, *1448, Biblioteca Apostolica Vaticana, Vatican, Rome.*

this transition, we must turn to the earlier medieval tradition of European cartography. As we do this, we are also finally about to launch into the speculative question of how van Eyck's lost map might have looked, and in order to do that, we must turn at last to more strictly art-historical evidence.

Fazio's description makes it clear that the lost map was not simply a functional map but was rather a splendid work, appropriate both to Alfonso I and to the prince of the Belgians. Thus, although it might have put places as nearly as possible in true, measurable order and might have been useful for the estimation of distances, it must also have been, as *mappamundi* had long been, a fundamentally symbolic map.

Charles Sterling, who located the making of the map in the context of Philip the Good's plans around 1436, offered two manuscript paintings as records of its appearance. Both show the Last Judgment. One, from the *Hours of Alfonso of Aragon* in the British Museum, painted in 1442, is very abbreviated, and the most important connection between the two for present purposes is the representation of the world as a projected disk. The second, much more informative image (figure 3.7), was made by the Spanish painter Juan de Cáceres between 1514 and 1524.[35] To be sure, this is also a condensed view of the world, but what might be recollections of ruined Eyckian structures and blue Eyckian mountains are also to be seen. Sterling fleshed out these resemblances by comparing them to the much grander landscape behind the *Madonna of the Chancellor Rolin* (figure 3.8), painted around 1435. The heads at the four intercardinal corners of the world might be the angels restraining the winds (*Revelations* 7:1), but these heads are puffing lustily, and such wind gods are a feature of Ptolemaic world maps (see figure 3.16).[36] As is appropriate to a Last Judgment, the continents and oceans give up their dead, as fish disgorge those they have swallowed, a detail Sterling compared to the disputed Metropolitan *Last Judgment* (figure 3.9) from the later 1430s. Sterling concluded that the lost map offered a bird's-eye view—that is, the world as a projected disk, realist landscapes, continuous and slightly curved, scattered with small images of fields and little "skylines" of cities like those to be seen in maps going back to antiquity, with roads, rivers, and chains of mountains.

Considered more closely, Juan de Cáceres's watery disk is a most extraordinary projection of what historians of cartography call a medieval OT map (or T-in-O map), which, whatever transitions, transformations, emendations, and adaptations it might have undergone was a diagrammatic representation of the Roman and then of the Christian world.[37] The Roman Empire crumbled, and as the Crusades themselves testify, this representation of the Christian world was more a hope than a reality, since, limited as it might be, that world—even

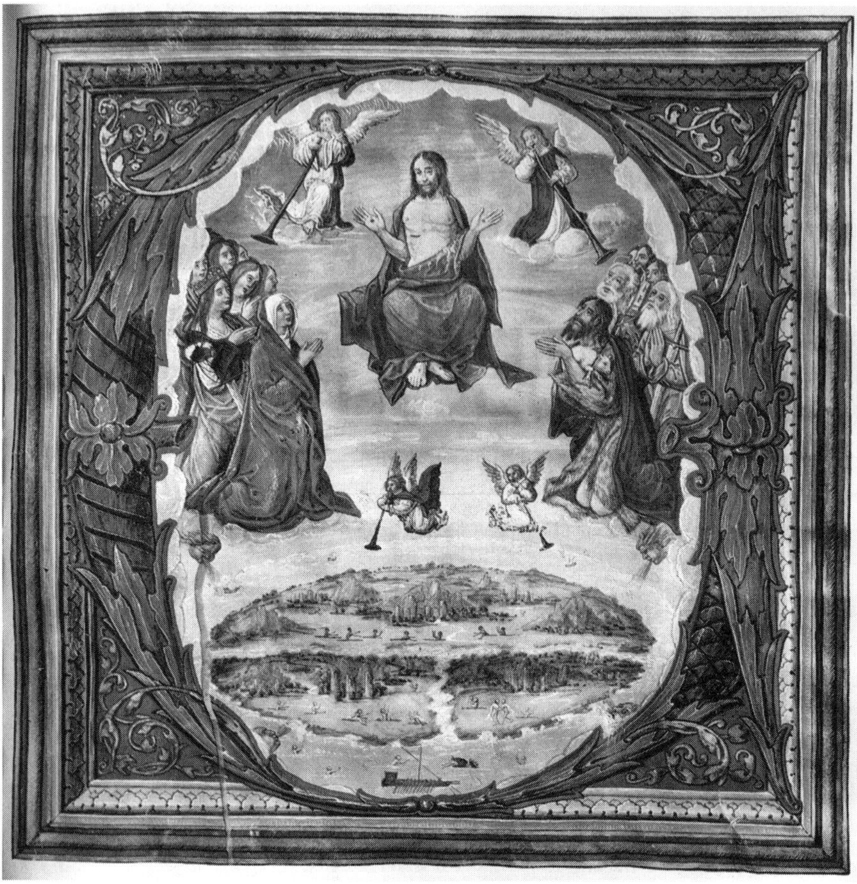

FIGURE 3.7. *Juan de Cáceres,* Last Judgment with Map of the World, *1514–24, antiphonary, Cathedral, Granada. From* Revue de l'Art *33 (1976): 73, fig. 111.*

Jerusalem, its center—was not exclusively Christian. The schematic map from a fifteenth-century edition of the seventh-century *Etymologies* of Isidore of Seville (figure 3.10), follows a pattern to be seen in any number of medieval manuscripts (and in images of God or Christ holding the orb of the world). In Jan van Eyck's lifetime, of course, there were still three known continents—Europe, Asia, and Africa—and this old scheme was invariably followed. In medieval OT maps, east and often Paradise are at the top, and the T could be compared to the *tau* configuration of the cross.[38] The T forms the variously identified waters of the earth, which empty into the great sea in the middle of the earth, the Mediterranean, as we still call it. These waters separate the three continents, divided up among the descendants of the sons of Noah, Shem, the

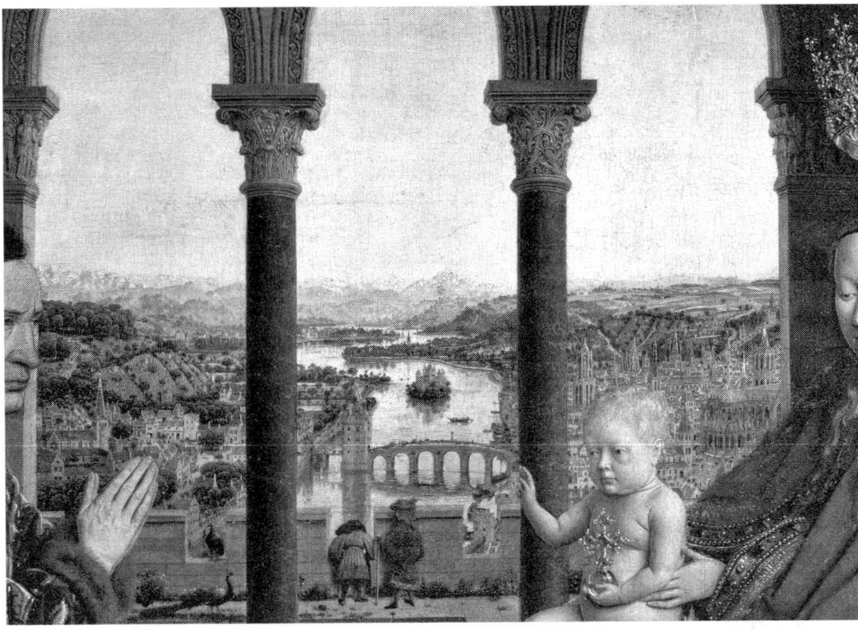

FIGURE 3.8. *Jan van Eyck,* Madonna of the Chancellor Rolin, *oil and tempera on wood, detail, landscape, ca. 1435, Louvre, Paris. Erich Lessing/Art Resource, New York.*

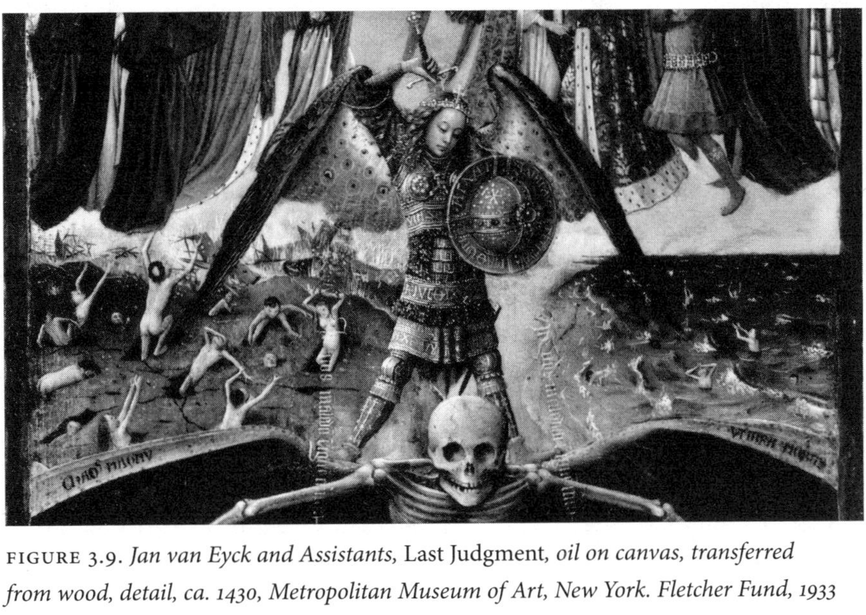

FIGURE 3.9. *Jan van Eyck and Assistants,* Last Judgment, *oil on canvas, transferred from wood, detail, ca. 1430, Metropolitan Museum of Art, New York. Fletcher Fund, 1933 (33.92ab), © Metropolitan Museum of Art.*

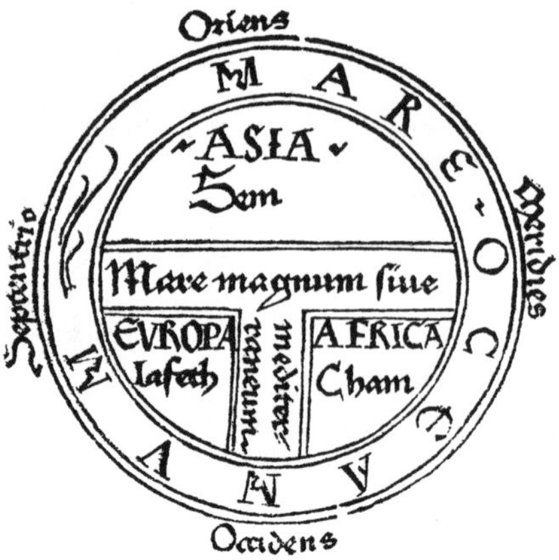

FIGURE 3.10. OT Map. *From Isidore of Seville,* Etymologiarum sive originum libri XX *(Augsburg: Zainer, 1472).*

eldest, who was given Asia, the biggest; Ham, who was assigned Africa; and Jepheth, who was given Europe. Jerusalem, *umbilicus mundi*, destination of Christian pilgrim and crusader alike, typically lay at the crossing of the T and the center of the O, the latter the surrounding Ocean Sea into which the Mediterranean emptied. Beyond the circle, Ocean Sea covered the globe of the world in ways about which no one had the least knowledge.

The thirteenth-century *Psalter Map* (figure 3.11) displays both the variability and the persistence of the OT scheme. Christ, censed by angels, is at the eastern top, holding an OT orb in his left hand, a sign of world dominion, with the rising sun and Paradise, from which five rivers flow, beneath him. Jerusalem is at the center. The waters of the world, in an elaborated T suggesting topographical description, join in the Mediterranean in the lower, deathly west, where dragons confront one another. The strangely deformed peoples at the edges of the world are shown far to the south,[39] to our right, and to our upper left, at another periphery, Gog and Magog are safely walled and gated.[40] Mountain ranges are shown in profile, and cities are marked by red triangles, sometimes clearly defined as structures and sometimes multiplied to stress their importance, as in the case of Rome.

By the early sixteenth century, when the *Last Judgment* in figure 3.7 was painted, the OT formula was decidedly old-fashioned; at the same time, the view of this ancient scheme of the world as an optical plane under a visual cone is most extraordinary. The three known continents are shown as if from a point of view high above to the west, and this three-dimensional projection of the traditional disk thus retains the ancient eschatological geography and translates it into then-modern terms.

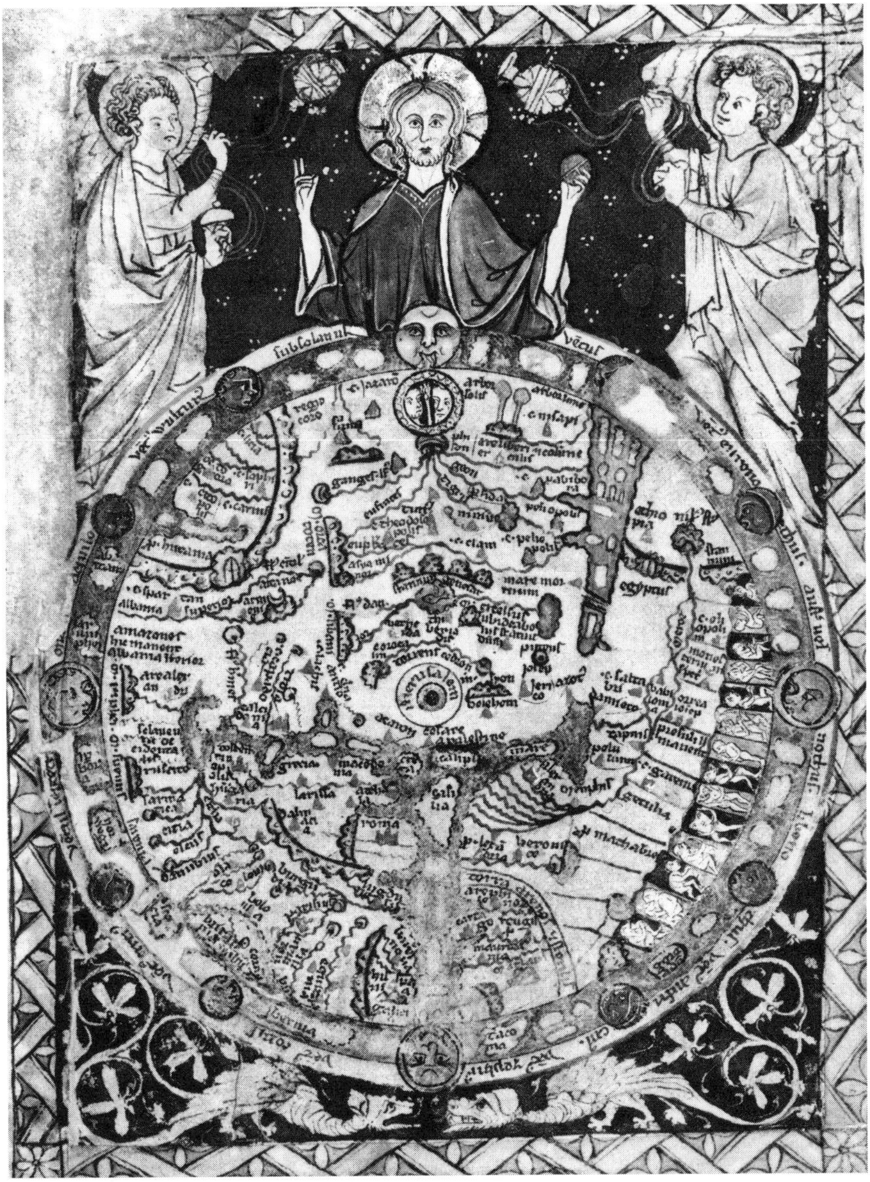

FIGURE 3.11. Psalter Map, *thirteenth century, British Library, London.*

In OT maps, east—and Paradise—is symbolically highest, but in Juan de Cáceres's bird's-eye OT map, where the three continents are seen as the base of a visual cone, east—and Paradise—is farthest away in the virtual space. We see the continents from a perch high above the western ocean—looking up the stem of the T, that is, through the Pillars of Hercules and into the Straits of Gibraltar, where the Mediterranean opens to the Atlantic. Now that the disk is tilted, we are *facing* the east, like Christians at the Last Judgment, when according to the Gospel of Matthew (24:27) "as the lightening cometh out of the east, and shineth even unto the west; so shall also the coming of the Son of man be." We are facing the east, and we are also facing the Holy Land at the crossing of the T. Beyond the Holy Land, between us and Paradise, is the mysterious remainder of Asia.

Another manuscript painting, playing another variation on the projected OT scheme, is much closer to Jan van Eyck in time, place, and provenance (figure 3.12). This is from a manuscript of Jean Mansel's *Fleur des histoires*, a history of the world commissioned by Philip the Good.[41] The many illustrations in this manuscript, painted in the 1450s, are attributed to the youthful northern French painter Simon Marmion. The inscription in figure 3.12 tells us that this is the world according to the *Etymologies* of Isidore of Seville, and in Asia, Shem points upward, apparently to the ark, beached atop the mountain behind him. Jerusalem (unlabeled) is to our right of Shem, presumably pushed aside to make Shem and Noah central. The Dome of the Rock is clearly recognizable in the representation of Jerusalem, which very much resembles that to be seen in another Eyckian painting, the *Three Marys at the Tomb of Christ* (figure 3.13), often attributed to Hubert van Eyck, who was dead by September 18, 1426. Whatever that connection may be, Marmion's map is once again not just an image of the world but of the cosmos. The continents stand for earth, rivers and ocean stand for water, then air is shown by clouds, fire by a vermilion band of meteors or comets, then finally the quintessence, the fixed stars.

Fazio described van Eyck's map as round (orbicular), which most *mappamundi* were. Since nothing was known of most of the world at any time the map might have been made, it seems unlikely that the map would have been a globe. At the same time, an impression of rotundity in Marmion's map is unavoidable, as mountains and cities in the distant east fade to white. The little hatchings on the waters of the world seem to stand for waves, and as they rise and recede they become smaller and closer together, so that by the time we get farthest to the east what had been the boundary of circling Ocean in older representations has become the distant edge of a sphere, and we seem to be gazing out over the curvature of the earth, an unknown Pacific horizon rimmed by clouds, aglow with

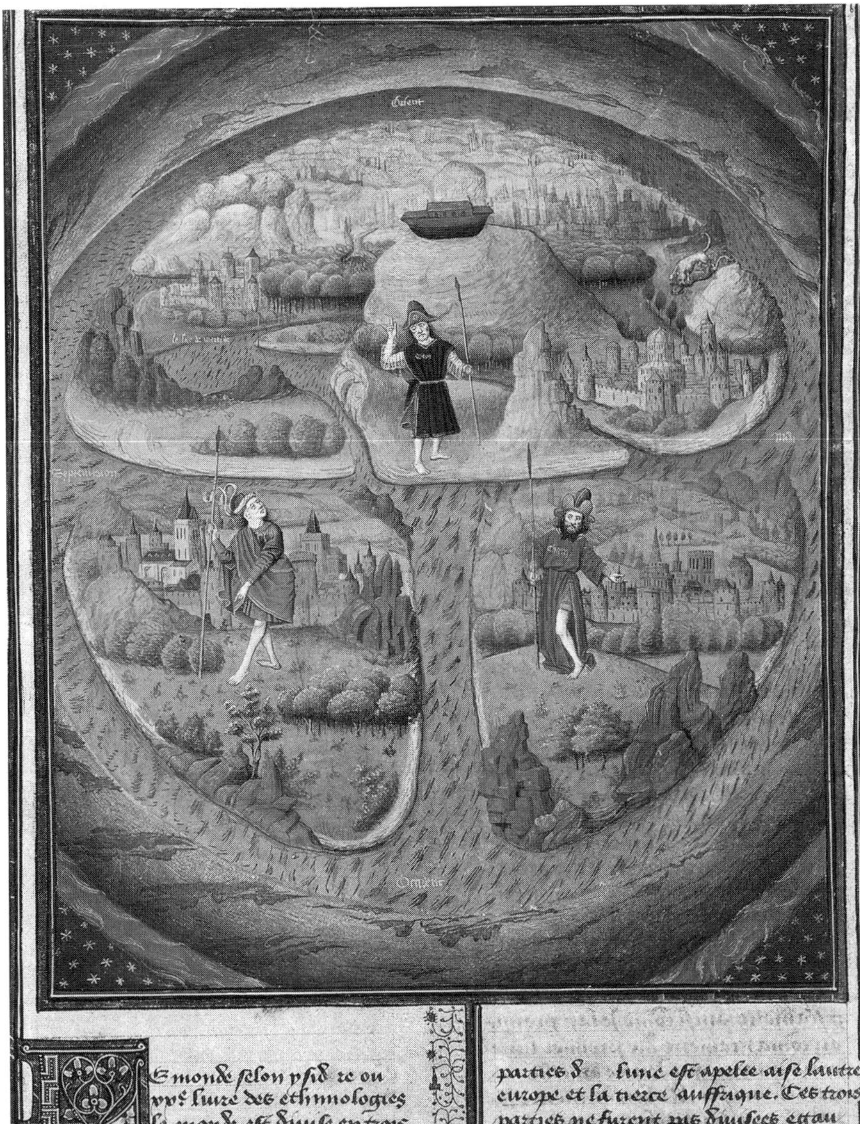

FIGURE 3.12. *Simon Marmion,* World according to Isidore of Seville, with the Sons of Noah. *From Jean Mansel,* Fleur des histoires *(ca. 1460). (Reproduced in color as plate 7.)*

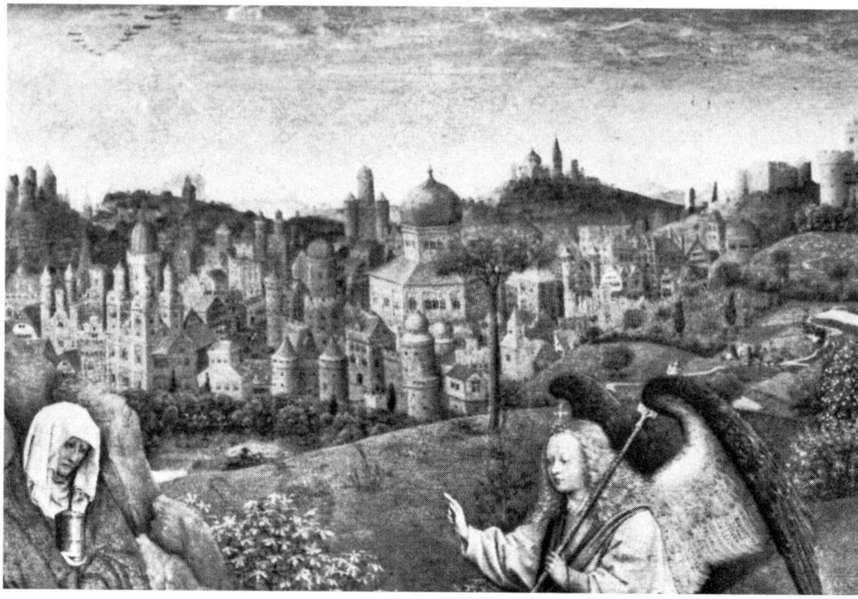

FIGURE 3.13. *From Hubert van Eyck,* Three Marys at the Tomb of Christ, *oil on panel, detail, ca. 1440, Museum Boymans van Beuningen, Rotterdam. Kavaler/Art Resource, New York.*

morning light. Asia may still be seen as the traditional OT upper semicircle, at the same time that it resembles the strange world aborning on the outside of Hieronymus Bosch's *Garden of Earthly Delights* (figure 3.14).[42]

If the maps by Juan de Cáceres and Simon Marmion (figures 3.7 and 3.12) are miniature reverberations and adaptations of Jan van Eyck's world map, then it may be concluded that the original was some variant of a projected OT map. Moreover, differences in style aside, there is a fairly close similarity in the disposition of mountain and cities in the two representations of Asia, perhaps suggesting a common source. The two projections, however, are quite different. Juan de Cáceres shows the world as a simple disk, but Marmion shows what might be called a semidisk; there is a split in projection from the lower to the upper half. Europe and Africa have the old shapes, which now, however, frame landscapes ending in distant profiles of cities and mountains. Asia remains semicircular, but the semicircle may also be seen as a hemisphere. Ocean follows the same pattern; the waves are seen from above but diminish toward the horizon. That is, the two lower continents are more nearly in plan (even if their shapes act as frames for deep virtual spaces), and the upper part is in projection. Although Marmion seems to have become impatient with painting

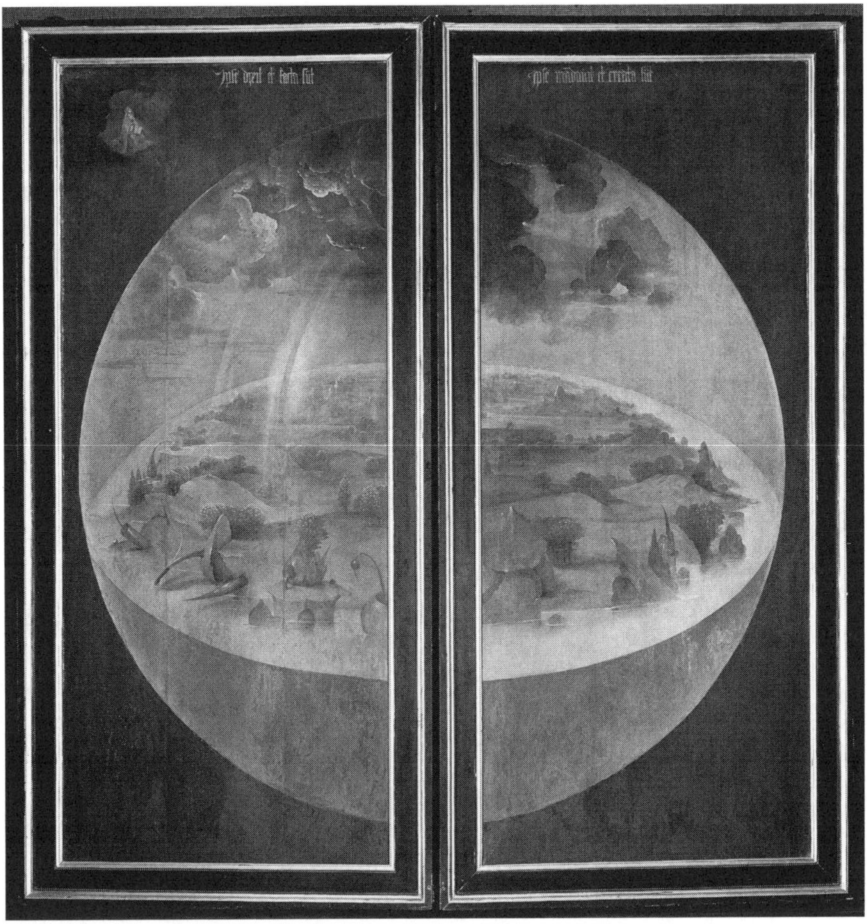

FIGURE 3.14. *Hieronymus Bosch,* Garden of Earthly Delights, *triptych, closed,* The Creation of the World, *ca. 1510–15, Prado, Madrid. Scala/Art Resource, New York.*

them, mountains and cities go on and on to the edge of the world and to Paradise (which, however, is not indicated).

I suggested earlier that Jan van Eyck's lost map might be related to the Vienna school of cartography and thus to transitional OT maps like that of Andreas Walsperger (figure 3.6). The claim is made in the legend of Walsperger's map that it is a "geometric description of the orb of the world made according to the *Cosmography* of Ptolemy, proportionally by longitude, latitude and climatic divisions." Walsperger's map, however, is not projected in the third dimension; on the contrary, like other transitional maps, it is entirely in plan. This would have made it easy to measure distances between places in the way Walsperger

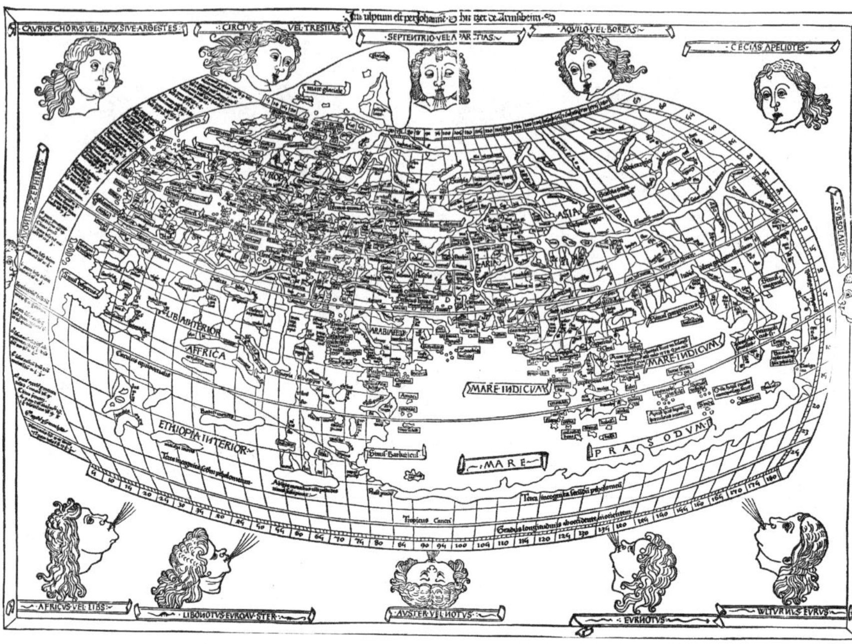

FIGURE 3.15. *Claudius Ptolemaeus,* Cosmographia *(Ulm, 1482), engraved by Johann Schnitzer of Arnsheim. Courtesy of Chapin Library of Rare Books, Williams College.*

instructed the map's user to do and as Fazio claimed was possible with van Eyck's map. But how can measurement have been possible on van Eyck's map if the disk was projected?

The presentation of the continents in Marmion's earlier miniature, which I have called a semidisk (figure 3.12), suggests an answer to this question. As we have seen, the Vienna cartographers attempted to reconcile OT maps and portolan maps with the new cartography of Ptolemy. Walsperger clearly used Ptolemy's tables but not his projections (figures 3.5 and 3.15). His *second* projection, which figure 3.15 shows in a later-fifteenth-century version, was most often used by Renaissance cartographers. In this projection, all the lines marking longitude and latitude are curved in response to the spherical surface they represent, and meridians diminish toward the North Pole. This grid is least distorted in the nearest part, nearest the division of the meridians along the equator. It would have been possible in these parts of the projection to construct a map on a fairly consistent grid, and even if the grid itself were not visible in the final result, it would have been possible to measure distances between places to a fairly high degree of accuracy.

In order to adapt Ptolemy's second projection to the traditional OT formula, it would first have been necessary to realign it from north to east, but since the earth is a sphere, the "pole," especially in a fundamentally symbolic map, might be easily shifted. In fact, to imagine an OT map aligned to the north is to imagine the denial of its eschatological meaning. If the alignment were changed and Europe and Africa were shown in plan, as in traditional OT maps and transitional maps, and Asia was projected from the latitude of Jerusalem northward, it would still have been possible to plot distances in the Holy Land. If the world consisted of three continents, most was known about two of them and virtually nothing was known of the farthest reaches of the largest of them.

Ptolemy not only offered a choice of projections but also divided mapping into two parts, geography and chorography. The general purpose of his geography was to establish metric relations between the habitable parts of the world (and the universe taken altogether), but his maps were also to show "larger towns and the great cities, the mountain ranges and the principal rivers," and "features worthy of special note on account of their beauty." Chorography is not concerned with size but with quality and true likeness, and so, Ptolemy said, the chorographer must be a painter. Ptolemy also said that the chorographer has no need of mathematics.[43] This of course does not mean that the chorographer need be ignorant of mathematics (and geometry), and, as Alberti's example might also show, by the time Jan van Eyck painted, the gap between painting and cartography was closing with the emergence of landscape.

Before finally trying to imagine van Eyck's map, we must consider more closely the question of alignment. As we have seen, Ptolemy's projections are all aligned to the north. Alignments in transitional maps were not fixed, however, and in figure 3.16, the map of the Venetian Fra Mauro of 1459—which is something over six feet in diameter—south is at the top, as it is in the Walsperger map (figure 3.6).[44] But in 1452, another Venetian mapmaker, Giovanni Leardo, followed the traditional OT pattern by placing east at the top, perhaps because he had also encircled the world with an elaborate Easter calendar (figure 3.17). (North and south, zones proclaimed too cold and too hot for habitation, are to our left and right.) At the top is Paradise, a large structure beneath a rubric cross, and Jerusalem, marked by a spire at the upper, eastern end of the Mediterranean, is at the map's center.

Judging from the miniatures of Juan de Cáceres and Simon Marmion (figures 3.7 and 3.12), Jan van Eyck's great map was a highly unusual version—a unique and unprecedented version—of a traditional OT map. Walsperger and Fra Mauro changed the traditional alignment, which Leardo retained; all three mapmakers continued to amend and amplify the old scheme as new, more accurate information came to hand, and it may be supposed that Jan van Eyck

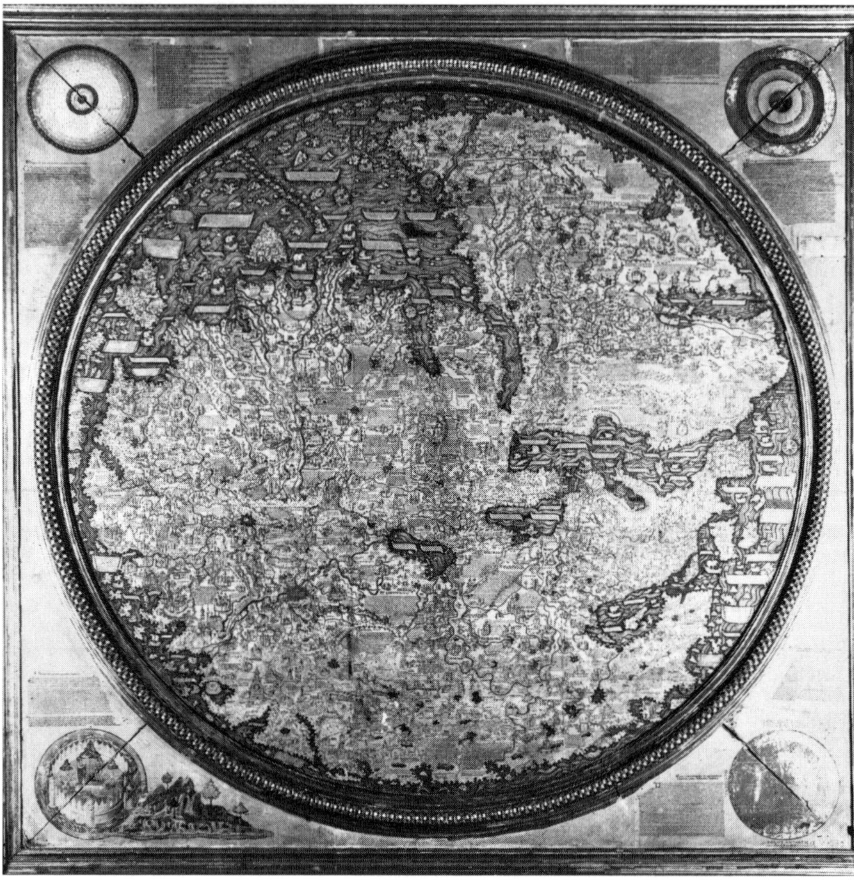

FIGURE 3.16. *Fra Mauro,* World Map, *copy of the map made for Alfonso V of Portugal in 1459 commissioned by the Signoria of Venice, Biblioteca Nazionale Marciana, Venice. Photo: Foto Toso.*

and his collaborators would have done the same. But it must be emphasized how extraordinary it would have been to *project* the traditional symbolic formula of the world. In clearly remaining an OT map while incorporating a projection like Ptolemy's, Jan van Eyck's map remained a symbolic, eschatological map, a vision of his patron's hoped-for Christian *oikumene*, as the association of figure 3.7 with the theme of the Last Judgment also suggests. At the same time, it would also have allowed distances to be estimated with accuracy in Europe, Africa, and the Holy Land.

Ptolemaic maps are keyed to latitudes and longitudes, defined in turn by the equator and the tropics, in their turn established by the relation of the earth and the sun, as are the meridians, one of which must be designated "prime."

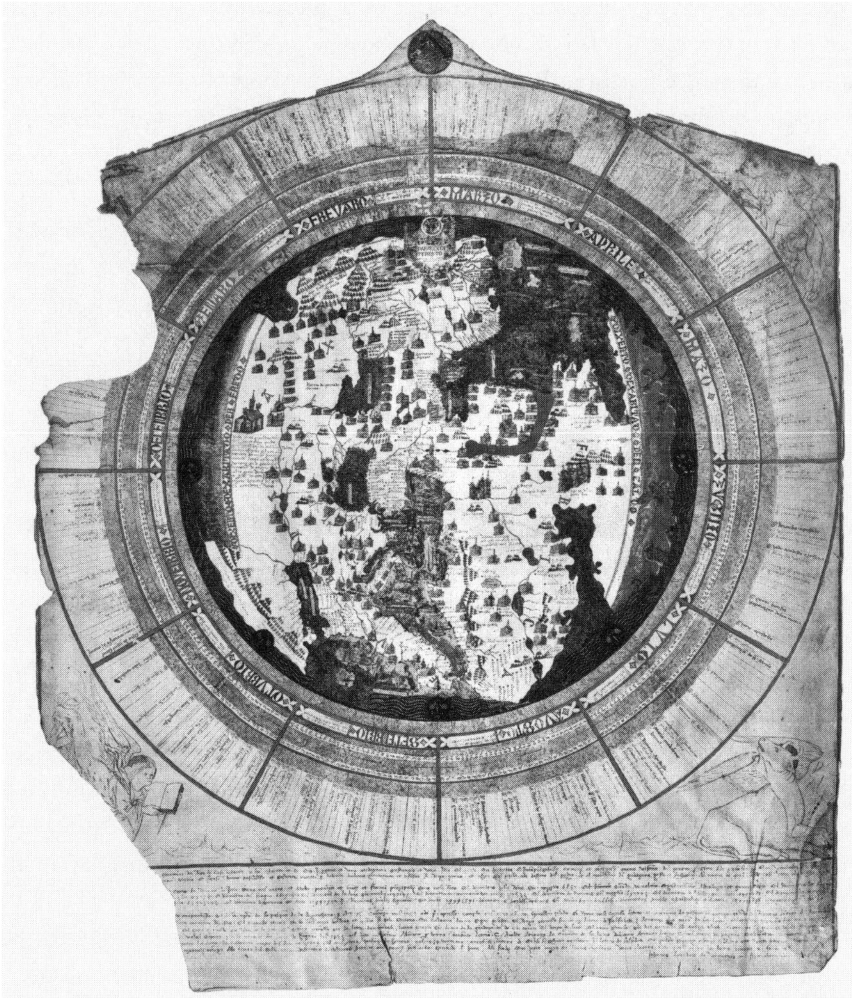

FIGURE 3.17. *Giovanni Leardo,* World Map, *Venice, 1452, American Geographical Society Library, University of Wisconsin–Milwaukee Libraries. (Reproduced in color as plate 8.)*

But a Ptolemaic scheme could in principle be applied with any axis, and if Jan van Eyck's symbolic Christian map placed the true pole of the world in the east rather than in the north, in Paradise rather than beyond Ultima Thule, it was the *oikumene* that was being mapped in either case. The north-south axis crossing the west-to-east axis at Jerusalem, the center of the world, defined the line at which the global surface of Asia began to turn and recede to the eastern edge of the world, finally across vast spaces of which in fact no distances were known.

If we return to the Walsperger map (figure 3.6) and rotate it ninety degrees to the right, so that east is at the top, and combine it in our imagination with the eastern edge of the world in Marmion's illumination (figure 3.12), we may be able to get some notion of van Eyck's map. Walsperger's map marks important places with mountains and buildings seen in elevation against the plan and projection of the map as a whole. And, as Sterling argued, we must imagine van Eyck's map filled with tiny, sparkling Eyckian profile paintings of major cities, mirrorlike rivers, and blue mountains. Largest among these cities would have been central Jerusalem, the entry to Asia, beginning a vista reaching through vast and necessarily fantastic reaches to Paradise. We might expect the visionary landscape over the curved horizon in the east to have been coupled with a Last Judgment or perhaps elaborated as the fountain of life and throne of the Trinitarian deity to be seen in other Eyckian paintings (figure 3.18).

I have summarized these arguments in figure 3.19, where the Walsperger map is shown in plan (as it was made) for Europe and Africa but in a modified Ptolemaic projection for Asia. Europe and Africa are thus describable in terms of a grid. If east rather than north is the "pole," however, then prime meridian and equator are reversed and the east-west "meridian" runs from Paradise to the Straits of Gibraltar through Jerusalem, where it is crossed by the north-south (our left to right) "equator" separating Asia from the other two continents, the old horizontal bar of the T. Division of this line would have provided measures for all three continents, curving meridians rising to Paradise, the grid defined by these divisions descending to Europe and Africa. In Marmion's map (figure 3.12), this boundary is curved, perhaps a clue to the beginning of a projection on a spherical surface, as in Ptolemy's second projection (figure 3.15). The known world is thus presented fully in plan, and there is no distortion of the Holy Land. Ocean Sea still encircles the whole, but the known world may also be seen to ride on the surface of the watery globe. To earth and water—continents and seas—should perhaps be added the spheres of the elements of air, with clouds and fire, as in figure 3.12.

Fazio tells us nothing about the size of van Eyck's map, which bears on the question of the degree of detail. Fra Mauro's large map is very detailed, and in

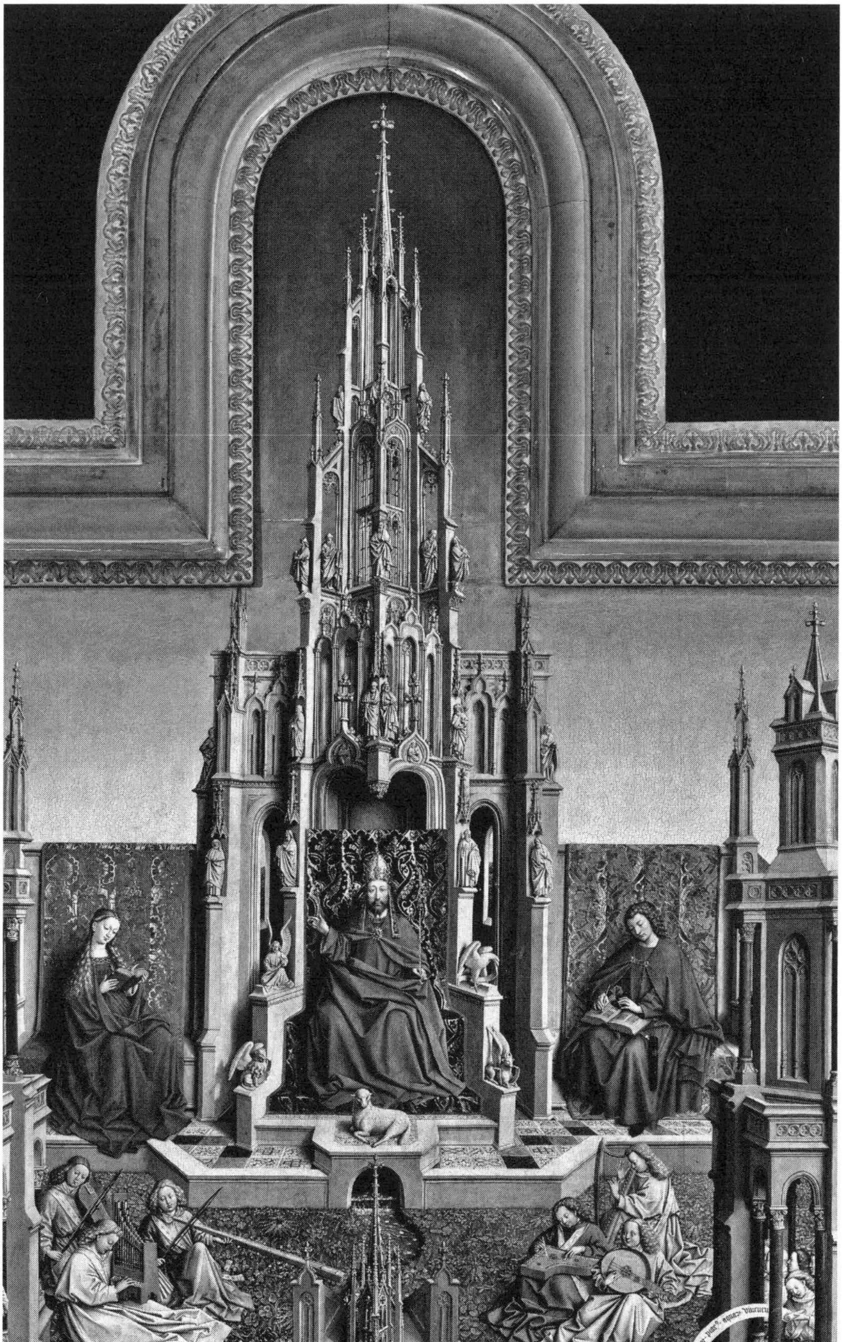

FIGURE 3.18. *Imitator of Jan van Eyck,* Fountain of Life, *oil on panel, detail, 1450s, Prado, Madrid.*

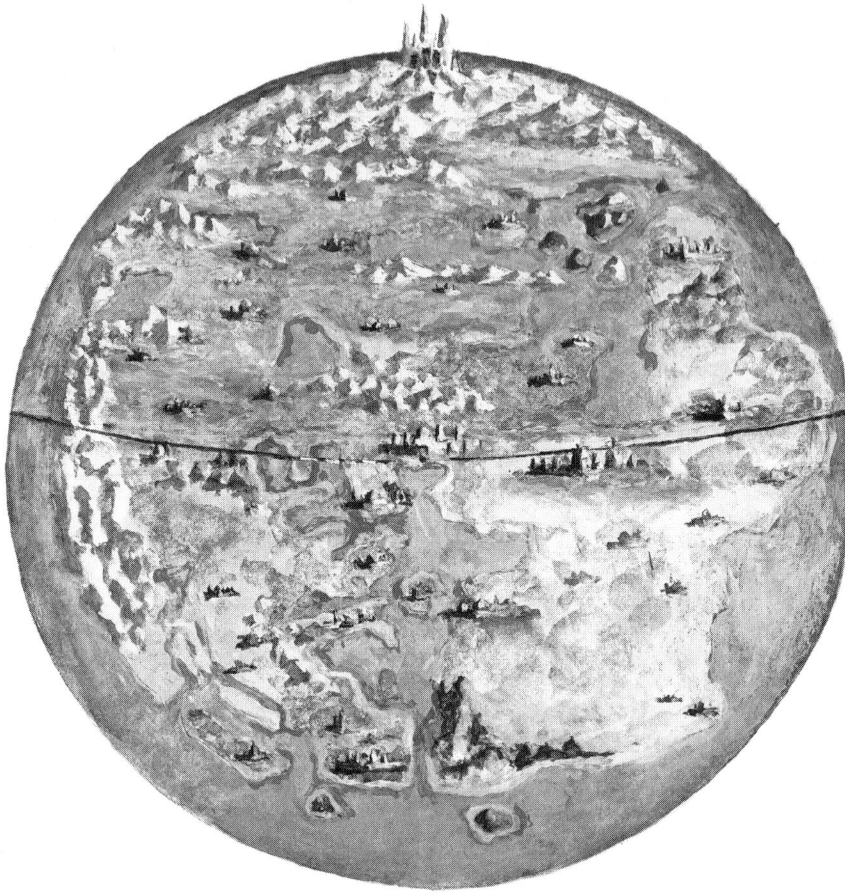

FIGURE 3.19. *Author's suggested reconstruction of Jan van Eyck's lost map. Cities are fewer and not to scale. There was probably a circle of air with clouds and perhaps a circle of fire to complete the four elements. (Reproduced in color as plate 9.)*

order to be useful in the ways its measurable distances suggest, van Eyck's map must have been fairly large, with major cities and mountain ranges in profile, chorography fusing with cartography in landscape as the world receded toward Paradise.

The projection of the world I have hypothesized for Jan van Eyck's lost *mappamundi* is finally not so different from the more properly microcosmic worlds of other Eyckian and pre-Eyckian works, the "plateau compositions" in which a near ground plane is set against a deeper space with precisely described horizon, as if we first looked down and then back (figure 3.20).[45] What this suggests is that the "mapping impulse," as Svetlana Alpers called it, resulted from

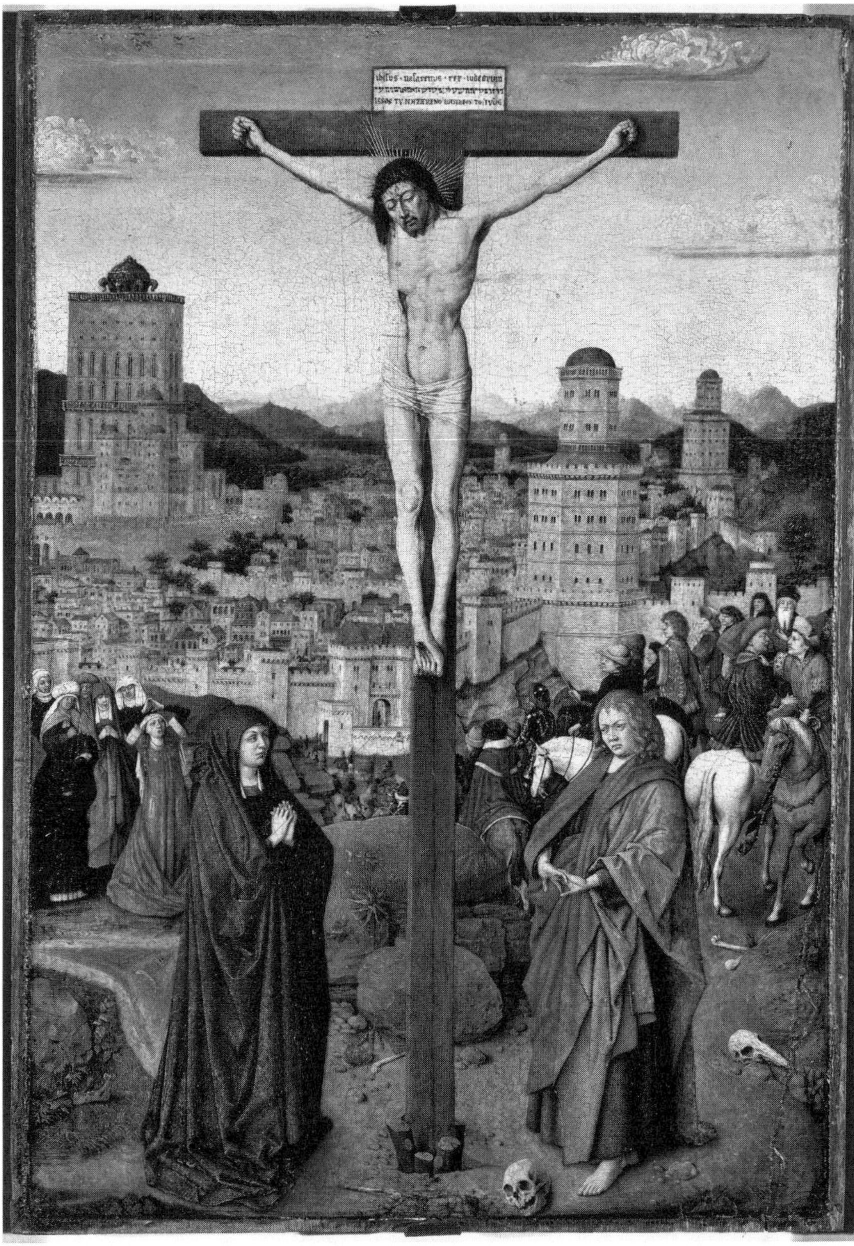

FIGURE 3.20. *Follower of Jan van Eyck,* Crucifixion, *oil on panel, Ca' d'Oro, Venice. Scala/Art Resource, New York.*

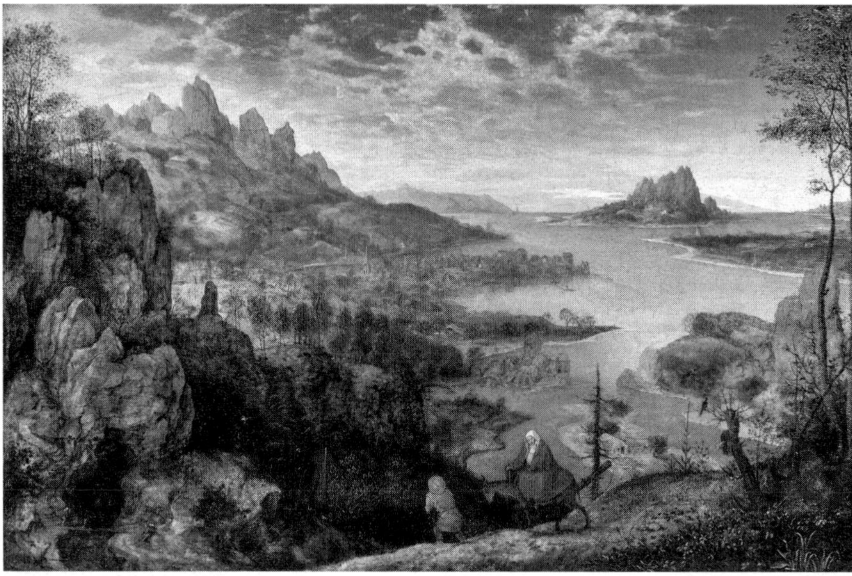

FIGURE 3.21. *Pieter Bruegel the Elder,* Landscape with Flight into Egypt, *oil on panel, 1563, Courtauld Institute of Art Gallery, London. Samuel Courtauld Trust, Courtauld Institute of Art Gallery, London.*

the codevelopment of cartography and "early Netherlandish painting."[46] The cartographic space of northern painting, as distinct from the urban space of Italian painting, also established the imaginative basis of the "world landscape" for later northern painting (figure 3.21).[47] And, to return to the theme of the last chapter, the world of the cartographic grid supplemented the metric world understood as the reflection in a plane mirror; landscape supplemented the near forms of Italian painting, giving it new subjective depth and finally, with Leonardo da Vinci, becoming not just aerial perspective but the intervening mystery of endlessly active matter. In the North, the "world landscape" became a foundation of modern imagination, a first taste of sublimity and of a certain vast silence, as it spread with European culture to provide the basis for new visions of dominion in parts of the world far from Europe itself of which Jan van Eyck could not have dreamed (figure 3.22).[48]

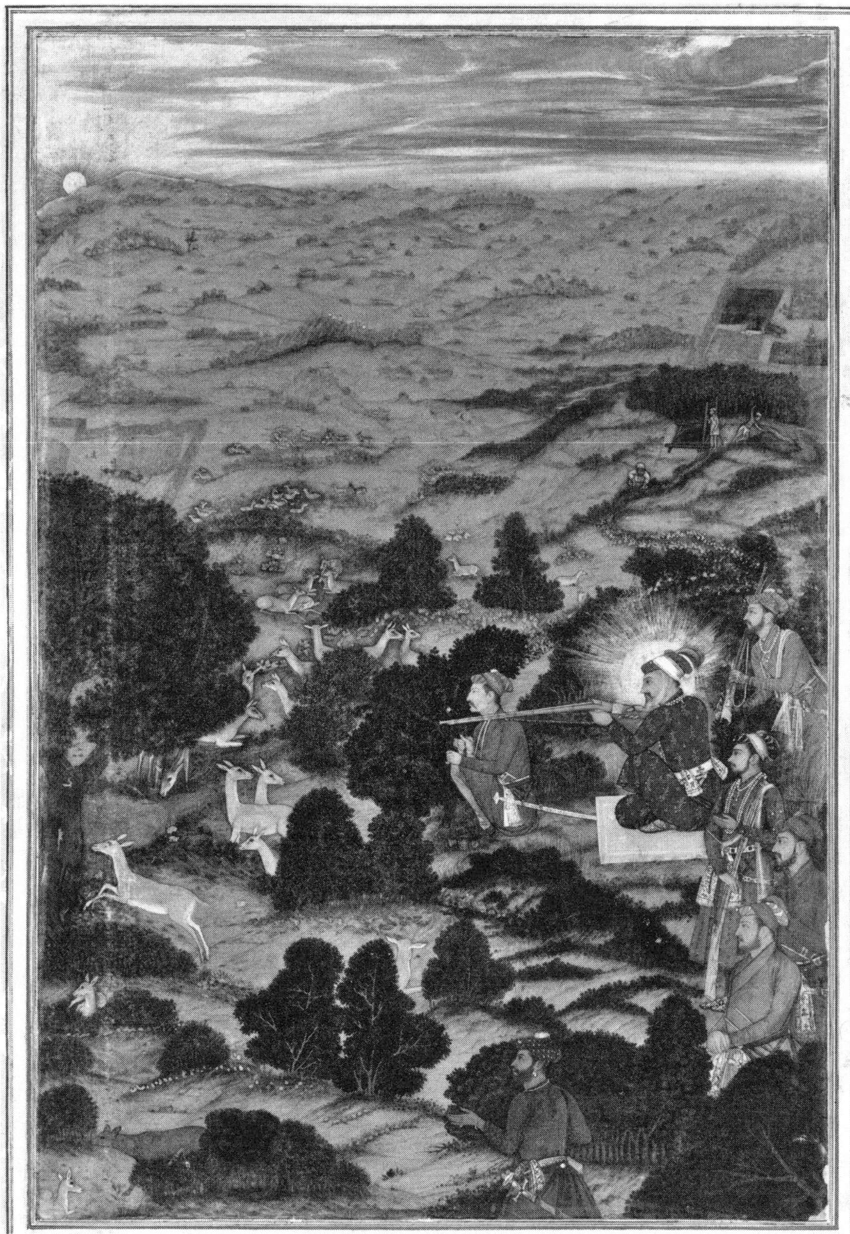

FIGURE 3.22. *Artist unknown,* Shah Jahan Hunting, *opaque color on gold-flecked paper, after 1634. From* Padshahnama, *f. 165a. The Royal Collection, © HM Queen Elizabeth II.*

CHAPTER 4 Narcissus and the Mirror without Stain

The map in figure 4.1 is upside down to the modern eye. The geography is familiar enough, however, and the map shows routes from cities in Germany to Rome. It is aligned to the south, and the peninsula at the center bottom is Denmark. Scotland is to our lower right, and the larger peninsula at the top is Italy, shown as far as Naples. The map was made by Erhard Etzlaub, who in addition to making maps was a surveyor and maker of pocket sundials. Etzlaub lived in Nuremberg, which is almost exactly at the map's center. This print, which was probably meant for the use of pilgrims in the jubilee year of 1500, combines the new technology of printing with a strain of earlier fifteenth-century cartography associated with Nicholas Cusanus,[1] the cardinal/theologian/philosopher best known as the author of the *De docta ignorantia*, a mystical-Platonic skeptic dialogue elaborating a vision that came to him in 1438 as he returned to Italy aboard ship from Constantinople.[2] Cartography came under the category of the relatively certain mathematical knowledge of which Cusanus argued the human mind is capable, but his interest in mapping might still have echoed the cartographic project of Roger Bacon and others in the thirteenth century discussed in the last chapter. However that might be, as his voyage itself suggests, Cusanus was closely involved in the negotiations culminating in the Councils of Ferrara and Florence, which, at least in hope and principle, effected the union of the western and eastern churches, transforming the *oikumene* of Ptolemy into an ecumenical Christianity. What was achieved in hope and principle, however, was not achieved in political fact, and Constantinople fell to Mahomet II just fifteen years after Cusanus's voyage.

But to return to Etzlaub's map (figure 4.1), it is drawn to scale, and the instructions for use at the bottom are roughly similar to those to be seen on the Walsperger world map (figure 3.6) discussed in the last chapter. On the Etzlaub map, "Mittag" is marked at the top, and perhaps noon in Nuremberg defined the meridian. With this map and one of Etzlaub's portable compass/sundials—which were called sundials for pilgrims—it might have been possible to find the way to Rome with a new confidence and sense of direction.

Although it was state of the art in 1500, more than half a century after the death of Jan van Eyck, no such map—and certainly no printed map—would have been available in late 1425 or early 1426, when van Eyck undertook his pil-

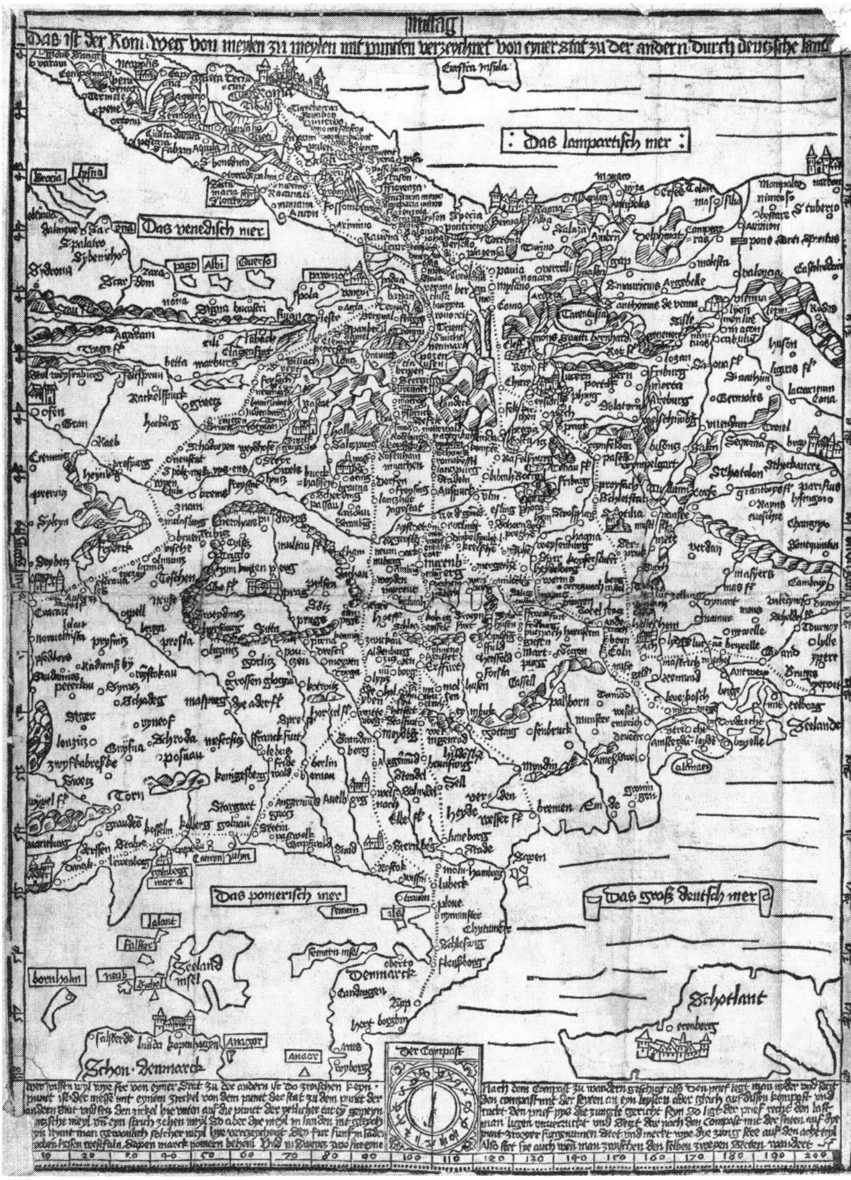

FIGURE 4.1. *Erhard Etzlaub,* Romweg, *woodcut broadsheet, ca. 1500, British Library, London.*

grimage for Philip the Good. Still, the routes were certainly older than the map that describes them. Bruges and Lille, where van Eyck lived by this time, are far to our lower right of the map, near the beginning of one of the routes measured off by circles and dots; this route begins at Nieuwpoort, where travelers from farther north must have disembarked, and leads to Zurich and Lucerne, through the Alps to Como, Milan, and Pavia. From there it goes to Bologna, then to Florence, Siena, and Rome. Someone going to the Holy Land in 1425 might have gone by other routes to Italy and by ship to the Levant. But if, as was usual, the pilgrim to the Holy Land also visited Rome, this might have meant a trip southward or northward along the route I have just briefly traced. And this route, as we have just seen, would have passed through Florence.

If he was on pilgrimage, van Eyck might be expected to have visited one of the great shrines of Europe, the miracle-working acheiropoetic *Annunciation* in the Church of the Santissima Annunziata in Florence, which was freely copied by painters throughout Italy and to which donations were made by the Duke and Duchess of Burgundy.[3] If, following on the premise of the last chapter, van Eyck had a working interest in cartography, he might also have had other reasons for visiting Florence. As we have seen, Florence was a major center of cartography, and Brunelleschi's friend Paolo Toscanelli, who had become the lifelong friend of Nicholas Cusanus while both were students at the University of Padua, had returned to Florence in 1423 or 1424. And once again, cartography, with its practical roots in surveying and applied astronomy, was also intimately related to the geometry of vision.

If Jan van Eyck passed through Florence in 1425 or 1426, he could have seen some of the finest works of the city's early Renaissance. Dates and sequences of Italian works themselves are of course disputed, but Brunelleschi's dome (figure 2.4) and Old Sacristy would have been well under way, Lorenzo Ghiberti's first Baptistery doors—in what can only have been a very congenial style for a Burgundian artist—would have been newly completed, Gentile da Fabriano's Strozzi Altarpiece in Santa Trinita would have been recently completed, as would Masaccio's *Sagra*, commemorating the consecration of Santa Maria del Carmine in 1422; the Brancacci Chapel in the same church would have been in medias res, and so might have been the *Trinity* in Santa Maria Novella (figure 2.6). Filippo Lippi, born in 1406, would have been about twenty years old. "Northern" features are often noted in Lippi's early work, he is sometimes assigned a "German" master, and if Vasari can be believed when he writes that Filippo decided to become a painter as he watched Masaccio at work in the Carmine,[4] then perhaps he also witnessed the visit of the great painter of the Duke of Burgundy.

Masaccio's older partner, Masolino (he was born in 1383 and so would have been in his early forties), was in Hungary with Pippo Spano from September 1425 to July 1427, and it might be supposed that during his absence, Masolino, who had begun the decoration of the Brancacci Chapel, painting the vaults and lunettes, passed the work on to Masaccio. If Jan van Eyck saw this work in progress, then there might be a historical basis for both the frequent comparisons and the contrasts between Masolino's *Temptation and Fall* (figure 4.2), Masaccio's *Expulsion from Paradise* (figure 4.3), and *Adam and Eve* of Jan van Eyck's Ghent Altarpiece (figure 4.4). Van Eyck might have learned from the two Florentine painters but might also have returned home with the intention of surpassing them in his own work.

Such resemblances and differences aside, a deeper and simpler similarity between the Brancacci Chapel and the Ghent Altarpiece may be noted—namely, the unification of natural and depicted light. The figures in the Brancacci Chapel are all modeled as if from a window in the altar wall, the altar itself, or a source of light behind the altar, the liturgical East. The Ghent altarpiece faces into its chapel from the east (figure 4.5). Figures on both the inside and the outside of the altarpiece, whatever their differences in scale might be, are lighted from our right, corresponding to the light from the windows in the south wall of the chapel. This light does not simply model forms; rather, its reflections and even refractions are meticulously described in a rich variety of textures of cloth, jewels, and precious metals. When the altarpiece is closed (figure 4.6), light apparently falling from the chapel window casts much the same shadows into the *Annunciation* from the frame, from Gabriel's near wing, and from the figures of Jodocus Vijdt and his wife. It is a soft, diffused light, like daylight. The general illumination of the altarpiece thus identifies the virtual space of the paintings with the space and light of the observer, the space and light from which it is seen. At the same time, this natural light itself has also fallen into the reserved space and time of church and chapel, and it is this transformed and heightened light that is the agent of the extraordinary luminosity and splendor of the surfaces in the further reserved virtual spaces of the altarpiece. The patrons, Gabriel, Mary, prophets, and sibyls share a common light.

In the *Annunciation*, there is another source of light we cannot see, entering through the trefoil windows behind Mary. This light passes through a flask of water on the window's ledge and makes shapes of light like candles entirely of flame on the dark wall behind the Virgin Mary (figure 4.7). If James Snyder was right and the space of the *Annunciation* is a solarium, a royal gallery high on the westworks of a northern cathedral,[5] then something like late afternoon light enters from the southwestern sky. This light, however, would be eastern,

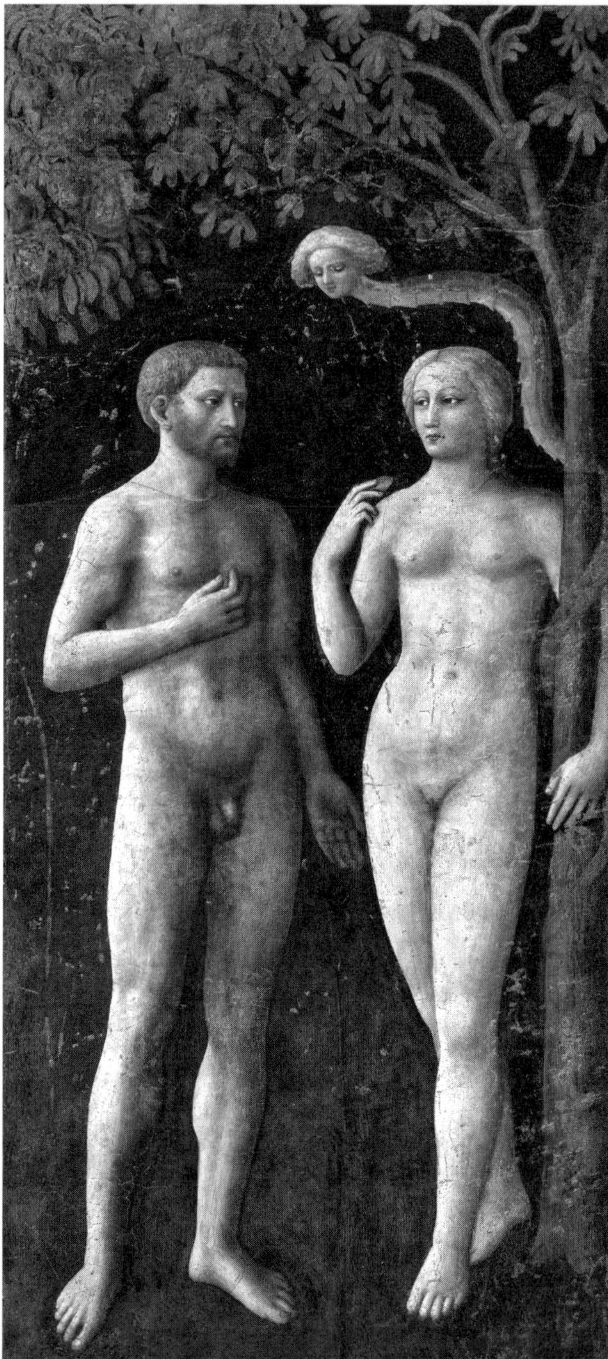

FIGURE 4.2.
Masolino, Temptation and Fall, *fresco, 1420s, Santa Maria del Carmine, Brancacci Chapel, Florence. Scala/Art Resource, New York.*

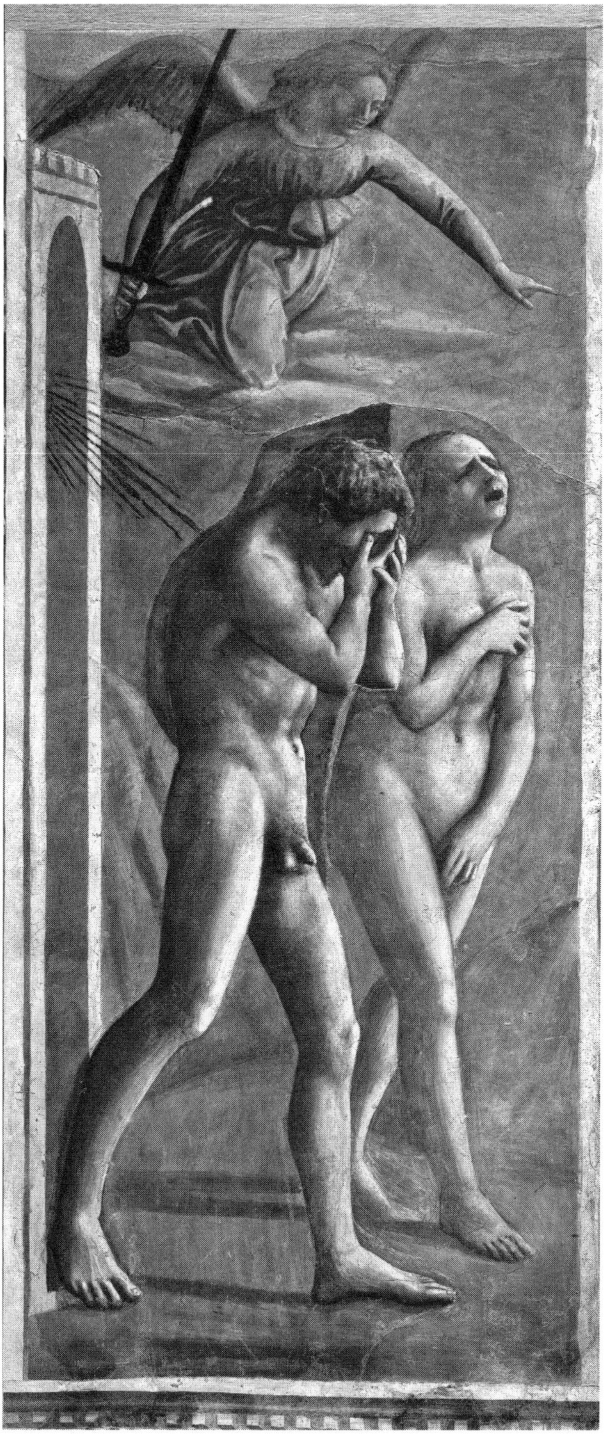

FIGURE 4.3. *Masaccio,* Expulsion from Paradise, *fresco, 1420s, Santa Maria del Carmine, Brancacci Chapel, Florence. Scala/ Art Resource, New York.*

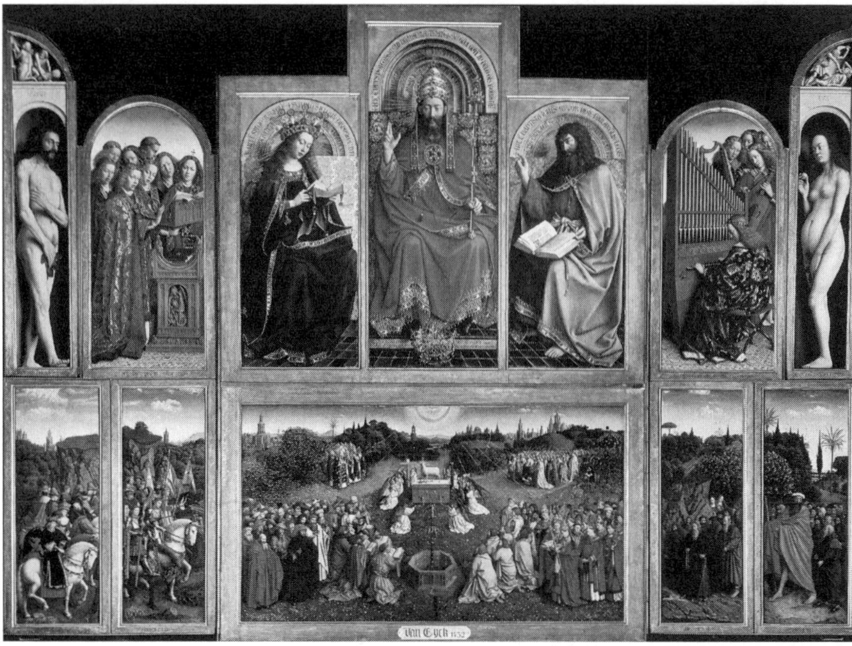

FIGURE 4.4. *Jan van Eyck,* Adam and Eve, *Ghent Altarpiece, open, completed 1432, Cathedral of St. Bavo, Ghent. Scala/Art Resource, New York.*

morning light in terms of the orientation of church and chapel, and it is clearly not simple sunlight that is entering; a half candle still catches the window's embrasure, as if a trinity of lights were about to be completed on the dark wall.

When the altarpiece is open (as it is shown in figures 4.4 and 4.5), reflections of the windows are to be seen in the heavenly accoutrements of the angelic chorus, and the same light, focused by the crystal scepter in his left hand, is to be seen on the raised right hand of the upper central figure of God the Father (figure 4.8).[6]

Technical examination of the *Annunciation* showed a major change in plans. Gabriel and Mary were first to have been in separate trefoil niches, like the figures of saints and patrons below them, before it was decided to show them in their present circumstances, a low chamber with tile floor and transversal beams above. As we are about to see, it has been argued that Alberti traveled in Northern Europe in the early 1430s, and this speculation fits well into the chronology of the Ghent Altarpiece, which was completed in 1432. But if van Eyck's new *Annunciation* in an interior suggests perspective, Alberti's treatise on painting was completed some years later, in 1435. Alberti, who was in Florence after 1428, may of course have been thinking about his ideas in the years be-

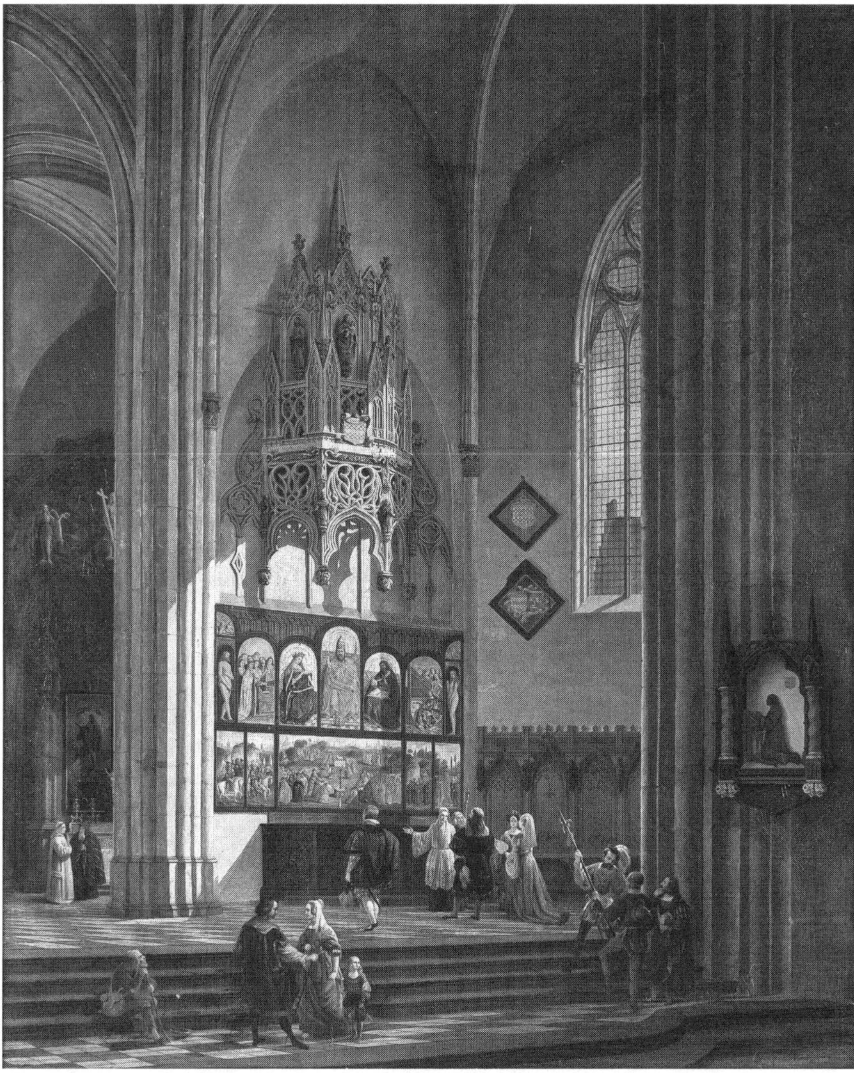

FIGURE 4.5. *P. F. de Noter,* Albrecht Dürer Visiting the Ghent Altarpiece, *oil on canvas, ca. 1840, Twenthe Museum, Enschede.*

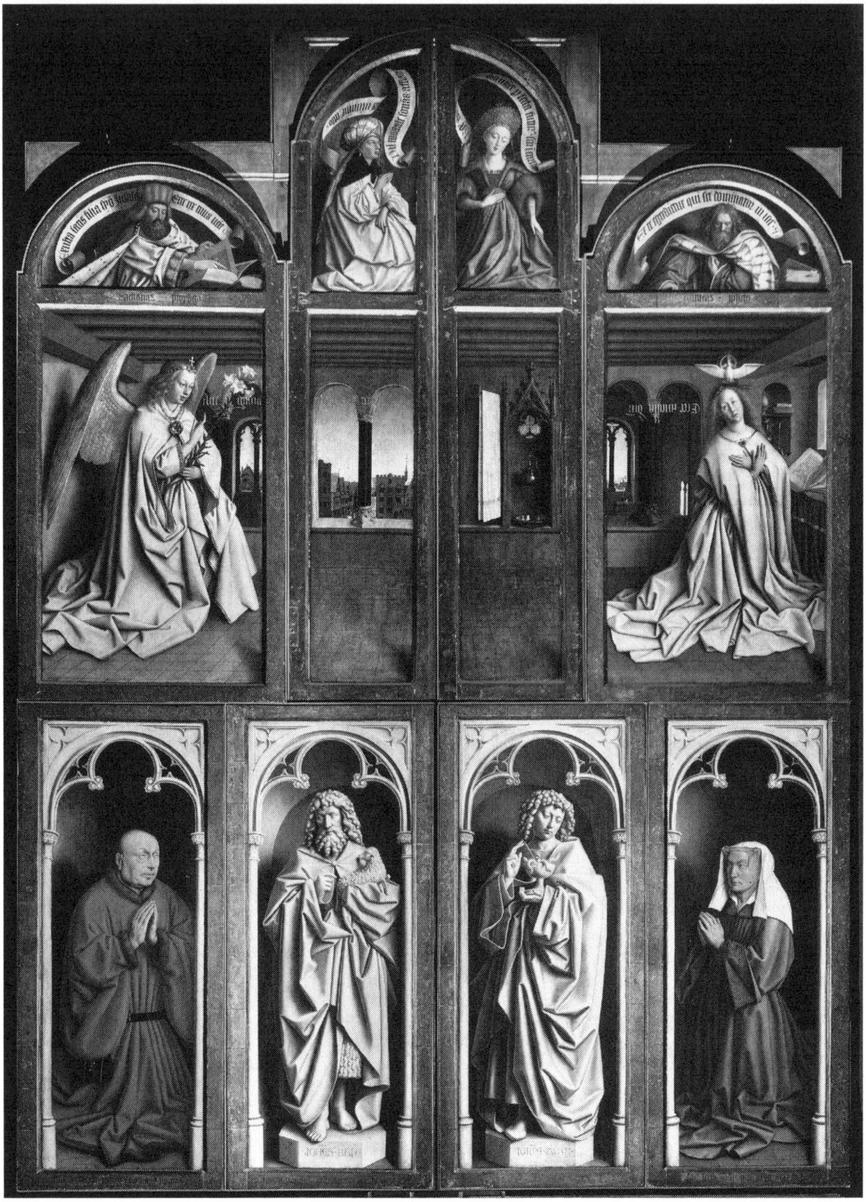

FIGURE 4.6. *Jan van Eyck, Ghent Altarpiece, closed, showing the Annunciation, prophets Zechariah and Micah, the Erythrean and Cumean sibyls, the donors, St. John the Baptist, and St. John the Evangelist, completed 1432, Cathedral of St. Bavo, Ghent. Scala/Art Resource, New York.*

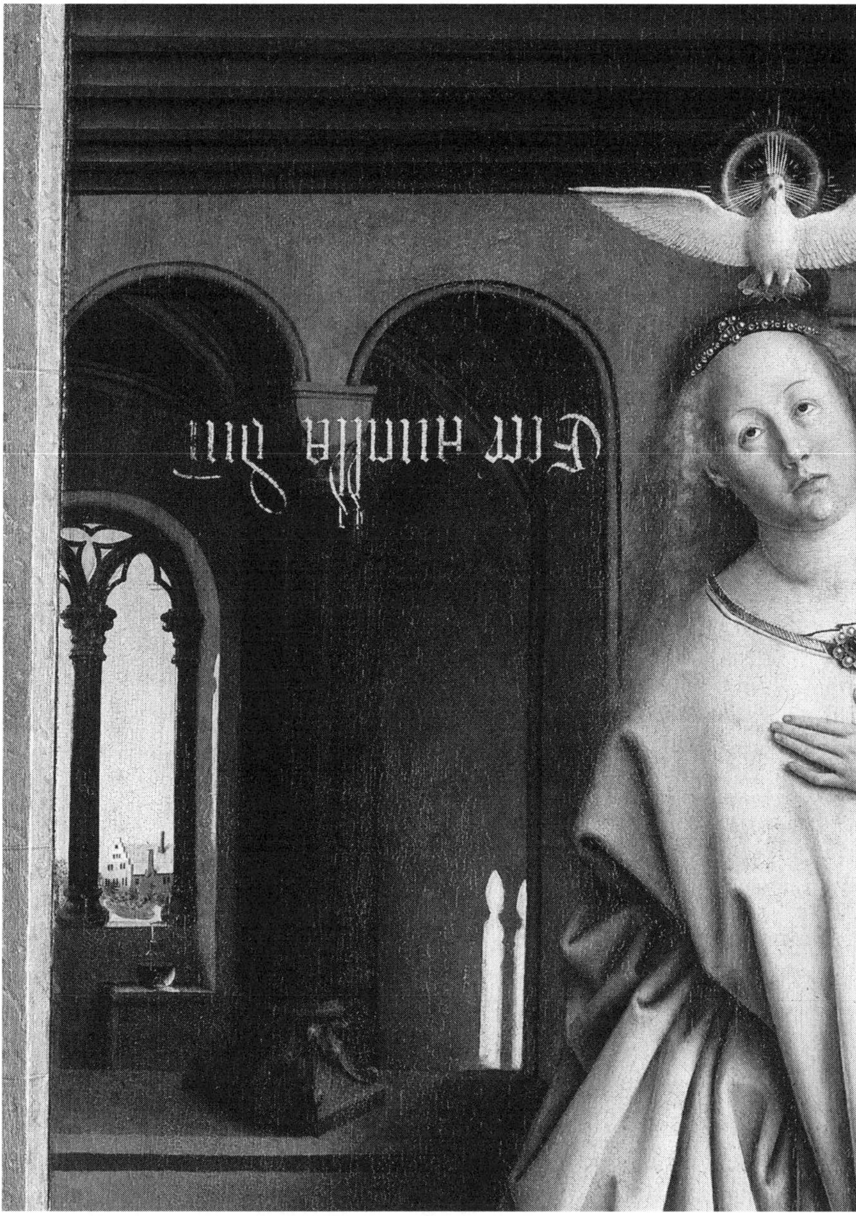

FIGURE 4.7. *Jan van Eyck, Ghent Altarpiece, closed, detail of the* Annunciation, *flask, and lights on the wall behind the Virgin Mary, completed 1432, Cathedral of St. Bavo, Ghent. Scala/Art Resource, New York. (Reproduced in color as plate 10.)*

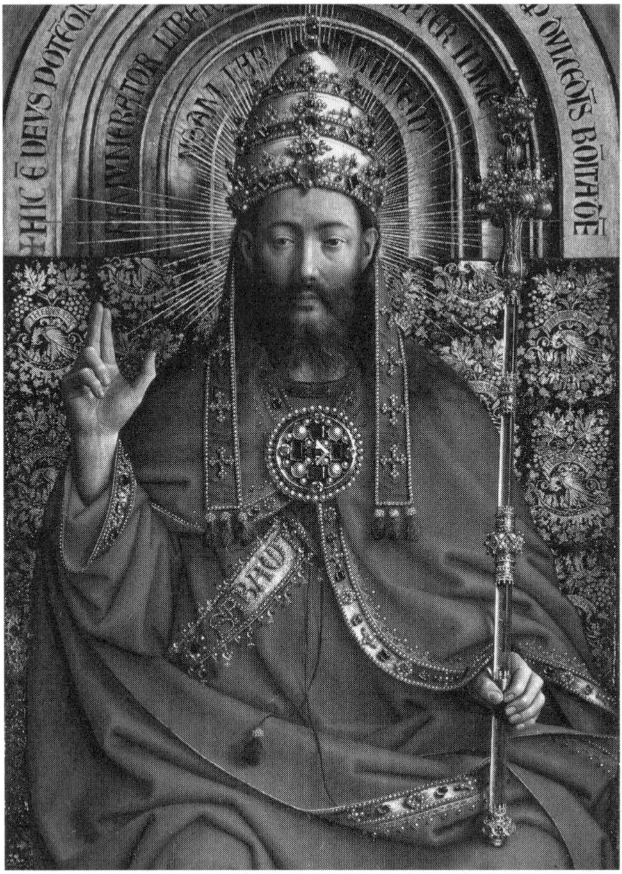

FIGURE 4.8. *Jan van Eyck, Ghent Altarpiece, open, detail,* God the Father with Crystal Scepter, *completed 1432, Cathedral of St. Bavo, Ghent. Scala/ Art Resource, New York.*

fore he committed them to paper, but this is not an Albertian pictorial space: the diagonals do not converge, and Alberti could not have praised the scale relation between the figures and the interior, even if this might have been the necessary consequence of the change in plan from niches to a single room. If, on the other hand, Jan van Eyck had already been to Italy five years previously, then his relation to events there might be much more complex—as they would have been in any case—than a possible brief encounter with the young Alberti would allow. Lotte Brand Philip compared the space in van Eyck's *Annunciation* to that of Donatello's *Feast of Herod* in Siena (figure 4.9), done around 1425, and to the *Annunciation* attributed to Masolino in San Clemente in Rome, which Vasari considered an early work of Masaccio. We no longer have the *Annunciation* with many columns drawn in perspective; the *St. Ivo* in a niche, seen from below with foreshortened feet; or the panel with life-size nude man and woman by Masaccio mentioned by Vasari.[7]

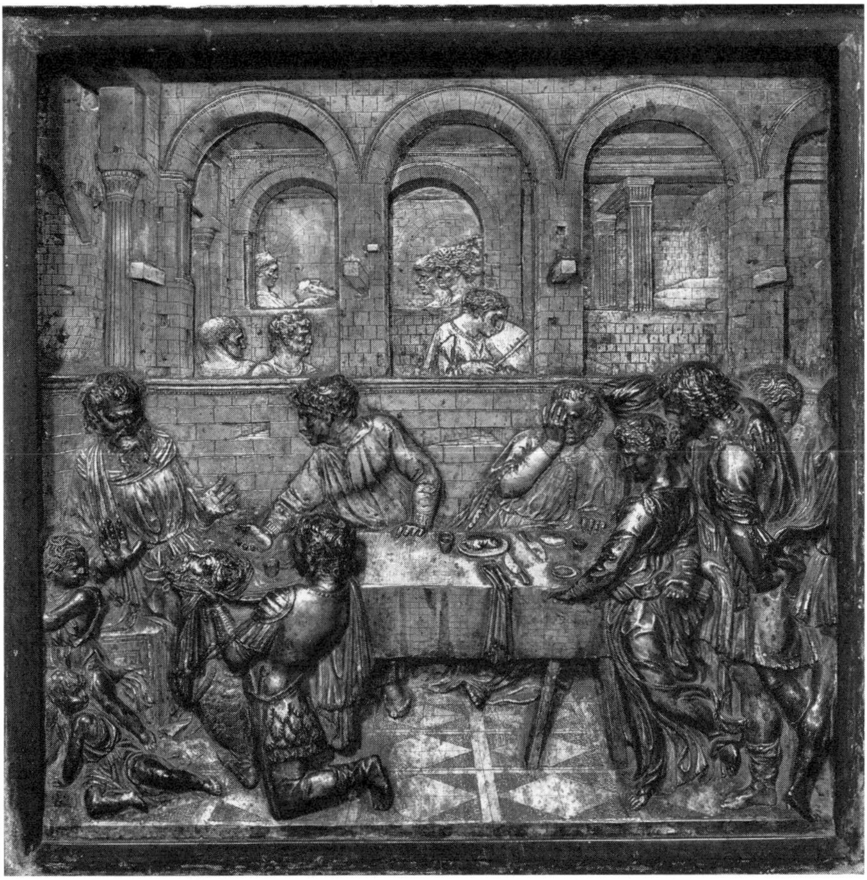

FIGURE 4.9. *Donatello,* Feast of Herod, *gilt bronze, 1425, Baptistery, Siena. Scala/Art Resource, New York.*

As the last paragraph suggests, there are enough tantalizing gaps and hints in the lives of Jan van Eyck and Alberti to have given art historians ample opportunity to imagine their having crossed paths. In the fourteenth century, the Alberti were deeply enough embroiled in Florentine political and factional strife to have been exiled from the city and thus probably excluded from a larger role in the early Florentine Renaissance. Alberti's grandfather was exiled from Florence in 1387 and his father in 1401. His father, Lorenzo, went to Genoa, where he met Bianca Fieschi, a beautiful young widow from one of the noblest Genoese Guelph families. The two did not marry, Bianca died in the plague of 1406, and so Battista, born in 1404, although a thoroughbred Guelph, was illegitimate and motherless, as was his brother, Carlo, born the previous year.[8] Lorenzo followed business interests to Venice in 1406, returning to Genoa briefly in 1408 to marry, with great public ceremony, Margarita di Piero Benini, the daughter of one of his alleged Florentine coconspirators.

By 1415, the precocious Battista (who added the name Leon years later) was in Padua, enrolled in the gymnasium of Gasparino Barzizza, where he received his humanist training. In 1421, at the age of seventeen, he was back in Bologna, studying canon and civil law at the university. His father died the same year, and at this point his fortunes began to worsen. The substantial inheritance his father had arranged for his two natural sons was challenged, and Battista's health broke, his doctors told him, under the stress of excessive study of the law. Worse afflictions followed, including loss of memory and a morbid visual sensitivity, and he gave up law for physics and mathematics (which must have included optics). Alberti may already have known Paolo Toscanelli from his years in Padua, and Toscanelli, whose friendships with Nicholas Cusanus and Brunelleschi I mentioned in the last chapter, was to be his lifelong friend.[9] Alberti had nonetheless completed his legal studies by the time the Florentine ban on his family was lifted in October 1428 (through the intervention of Martin V). Free to return to the city of his family, Alberti was also free to participate in its cultural life.[10]

Alberti was an early champion of vernacular literature, and in 1433, at about the same time he must also have been writing his treatise *On Painting*, Alberti began to write *Della famiglia*. Alberti wrote partly in order to master the Tuscan dialect, however, and as Michael Baxandall observed, *Della famiglia* must have been based largely on the recollections of older family members in exile.[11] His adult relations to his family were difficult and tense, with rumored threats of murder. Nonetheless, Alberti began his dedication to Brunelleschi with a parallel rejuvenation of Nature, "the mistress of things," and his newly returned family. He thought Nature had grown old and weary (*antica e stracca*), no longer able to give birth to those like the ancients. "But then I returned from the long exile in which we Alberti have grown old [*invecchiati*], to this our native country, adorned above all others," to find Brunelleschi, Donatello, Ghiberti, Luca della Robbia, and Masaccio. In any case, perhaps still in poor health and certainly without money, Alberti turned to the Church, which, in view of his good breeding and demonstrated brilliance, responded with a position and finally with a benefice.[12] It is thus between late 1428 and 1432, when he joined the papal court in Rome, that Alberti's speculative biography might intersect with the speculative biography of Jan van Eyck and that Alberti's journey to the north has been argued to have taken place.[13]

It was customary for bright young scholars to attach themselves to older princes of the Church. Tommaso Parentucelli da Sarzanna, for example, the future Pope Nicholas V (who as a young scholar recommended the ideal library to Cosimo de' Medici), was secretary to the Carthusian archbishop of Bologna, Niccolò Albergati[14] (in honor of whom he took his name as pope), and Alberti has been surmised to have had a similar connection. Albergati (c. 1375–1443) became archbishop of Bologna in 1417, became cardinal Santa Croce in Gierusalemme in 1426, and was nearly elected pope in 1431. (St. Antoninus, the archbishop of Florence, called Albergati himself a "mirror without stain," and he was beatified in 1744.)[15] Albergati spent his life in what might now be called church shuttle diplomacy. After the election of Eugenius IV in 1431, he was appointed to preside over the Council of Basel, which would remain a center of bitter contention for two decades. Albergati crossed the Alps in May 1431, traveling to Burgundy, Champagne, and Picardy, and it has often been suggested that Alberti was a member of Albergati's entourage.[16] Albergati stayed at the Charterhouse in Bruges from December 8 to 11, 1431, and in these days Jan van Eyck could have made the very beautiful and sympathetic silverpoint drawing (figure 4.10), with careful color notations, presumably for the painting (figure 4.11), which must have been done after Albergati's brief visit.

One would like to believe the sitter for the two portraits to have been Albergati and perhaps even to imagine Alberti looking over van Eyck's shoulder as he worked, but there are problems. Saintly as the subject may look, doubts have been repeatedly raised regarding the seventeenth-century identification, presumably based on an inscription on a lost frame. Albergati would have been in his mid-fifties in 1431, which might fit well enough, but it has been pointed out that both his habit and his haircut are wrong.[17] And if Alberti accompanied Albergati at the beginning of his journey, he had returned to Rome by October 1432, by which time he had become the secretary of Biagio Molin, patriarch of Grado, later of Jerusalem, at whose insistence he became a member of the College of Pontifical Abbreviators, responsible for the editing and publication of papal documents. Meanwhile, Albergati, after serving as papal legate to the Council of Constance in 1432, returned to Italy in 1433.

At the root of the Church's diplomatic efforts lay the desire—indeed, the urgency—of unifying Christian Europe, which was, like the Church itself, everywhere torn by conflict as the capital of Eastern Christendom succumbed to the Ottoman Turks. Albergati's travels on behalf of the Church took him to the Congress of Arras in 1435, where, although he did not succeed in ending the Hundred Years' War, he did at least achieve the reconciliation of Burgundy and France. A letter on the table in the disputed little painting of *St. Jerome in His Study* (figure 4.12), now in Detroit, is addressed to "Jerome . . . Cardinal priest of the church of Sta. Croce in Gierusalemme." Panofsky argued (even though the

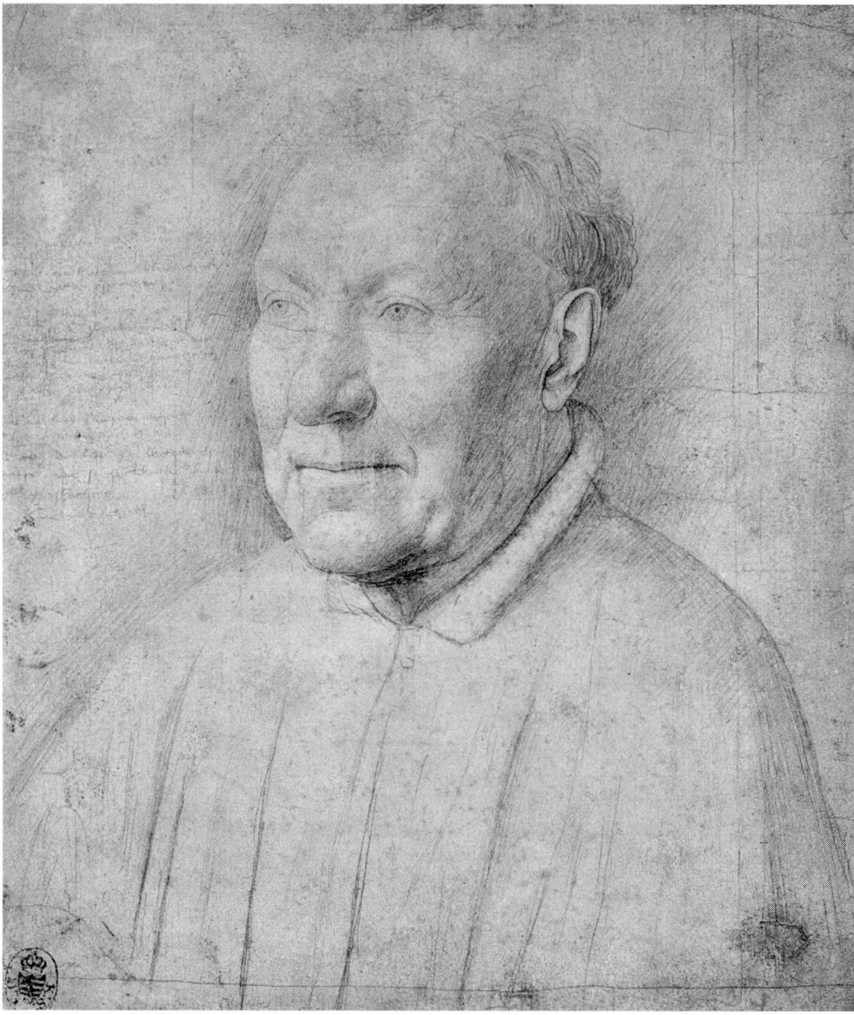

FIGURE 4.10. *Jan van Eyck,* Niccolò Albergati, *silverpoint, 1431, Kupferstichkabinett, Staatliche Kunstsammlungen, Dresden. Erich Lessing/Art Resource, New York.*

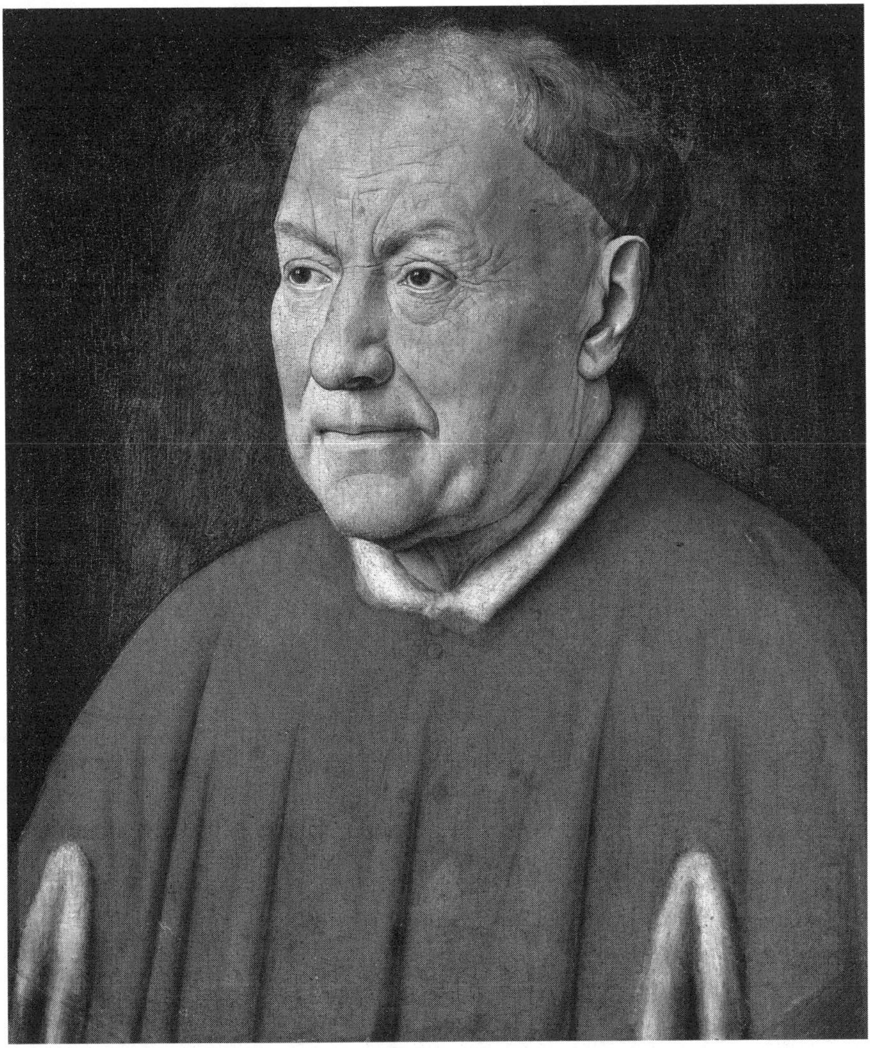

FIGURE 4.11. *Jan van Eyck,* Niccolò Albergati, *oil on oak, after 1431, Kunsthistorisches Museum, Vienna. Erich Lessing/Art Resource, New York.*

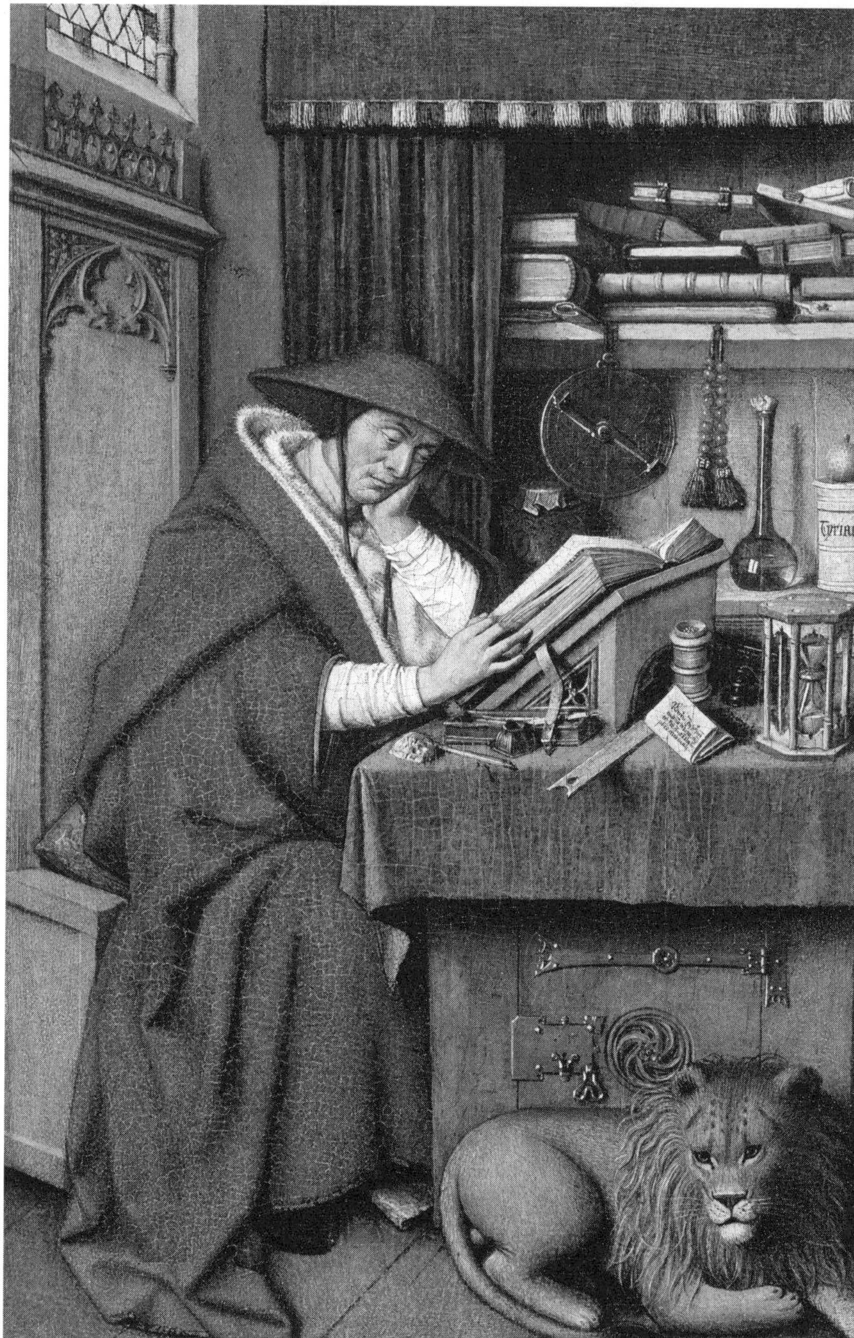

FIGURE 4.12. *Jan van Eyck,* St. Jerome in His Study, *oil on panel, ca. 1440, Detroit Institute of Arts. City of Detroit Purchase. Photo © 1984 The Detroit Institute of Arts.*

painting bears the date 1442, the year after van Eyck's death) that this painting, since it refers to Albergati's titular church, should be connected with his efforts at the Congress of Arras and even that it is a portrait of Albergati himself—he looks considerably younger than he did in the supposed portrait of 1431—as the great translator of the Vulgate, the ideal Christian humanist. Panofsky argued that Albergati commissioned the painting from Jan van Eyck; others argue that Philip the Good gave it as a flattering gift to the visiting papal representative,[18] perhaps hoping that he would heal Christendom, thus to clear the way for the Crusade of which the duke dreamed. Alberti, who returned to Florence from Rome in 1434, when Eugenius IV was run out of the Eternal City, finished *On Painting* in 1435 and went with the pope to Bologna in 1436. Albergati was a central figure in the Councils of Ferrara and Florence in 1438 and 1439, and if he brought this painting back to Italy, it may have entered the Medici collections after Albergati's death in May 1443.[19] All of these momentous events would have been known to Alberti and were no doubt of keen interest to him, but there are no indications of contact between Alberti and Jan van Eyck.

Alberti's biographers have felt more secure in their conjectures about possible travel to the north because of references he makes some years later in his *Ten Books of Architecture*, which he presented to his old friend Nicholas V around 1450. There he writes of the lime "I have seen" used by architects in Gaul, of the white stone used for roofing in Belgium, of how Germany "glistens with tiles," and of how German youths on ice skates "glide over the surface with a speed not matched by the swiftest birds."[20] Such details, if slight and distant memories, even perhaps examples of rhetorical *enargeia*, also seem uncontrived, quite different from his references to ancient authors. Taken at face value, these recollections might point not only to early travels in France, Germany, and the Low Countries but also to an early close interest in architecture of many kinds.

To return to one of the themes of the last chapter, the likelihood of Alberti's youthful travels might be increased by a brief review of his family's generations-old commercial activities. The Alberti traded in English wool and French fabrics; according to his biographer, Girolamo Mancini, they had workshops in London, Bruges, Ghent, Brussels, Cologne, and other cities in France. Their goods went from Paris to Avignon, then to Marseilles and Nice, where they were shipped to Pisa and Florence. There were also Alberti workshops in Greece, Syria, Spain, Hungary, and Scandinavia, and the house had connections in most Mediterranean ports.[21] In short, the Alberti, although banished from Florence before 1428, had been in one or another kind of contact with much of the known world, and Battista could have had many homes away from home.

In addition to the world map discussed in the last chapter, Bartolommeo Fazio described two paintings by Jan van Eyck. The first, it will be recalled, was a triptych, an *Annunciation* flanked by St. John the Baptist on the one hand and St. Jerome on the other. The triptych was originally commissioned by Battista Lomellini, and the outside of the flanking wings showed Lomellini and his wife in a room into which a ray of sun shone, perhaps with a significance related to that of the rays of light in the *Annunciation* of the Ghent Altarpiece (see again figure 4.7). The Lomellini were a Genoese merchant family closely allied with the Alberti, and when Leon Battista Alberti's grandfather fled from Florence to Genoa in 1387, he was taken in by them.[22] The Lomellini traded in Bruges, and Battista Lomellini and perhaps his wife might be supposed to have been in Bruges when van Eyck painted their portraits.

The third and last of the paintings Fazio described showed "women of uncommon beauty stepping from the bath" and belonged to a member of another major Italian court, Ottaviano della Carda, nephew and counselor to Federico da Montefeltro, the Duke of Urbino. Although this was no doubt a painting for male delectation, the women were modest, "their intimate parts veiled in fine linen; one, her face and breast uncovered, had her back reflected in a mirror on the wall behind her," like the mirror in the *Arnolfini Portrait* (figure 3.1). "Almost nothing," Fazio says, "is more wonderful in this work than the mirror painted in the picture, in which you see whatever is represented [in the painting] as in a real mirror."[23] Over and above the delights of its main subject, the painting seems to have been a concentrated display of *parerga*, of tour de force illusionism. "In the same picture there is a lantern in the bath chamber, just like one lit, and an old woman seemingly sweating, a puppy lapping up water." Of greatest interest in the present context is the landscape, seen through a window, in which were to be seen "horses, minute figures of men, mountains, groves, hamlets and castles, carried out with such skill you would believe one was fifty miles from the other." This last part of Fazio's description bears close attention and brings us back to Leon Battista Alberti.

Alberti's own interest in cartography yielded the *Descriptio urbis Romae*, a map centered on the Roman Capitoline, which he made in the early 1440s. This map may have been an ongoing project, and what he called his "miracles of painting"[24] were done in Rome some ten years earlier, before he finished his treatise on painting. Martin Kemp has written of these "miracles" as first in the unwritten history of artificial magic, such as peep shows and anamorphoses, in which carefully controlled conditions of viewing produce an astonishing effect of actual appearance.[25] These appearances were deep landscapes. As Kemp also observes, Alberti tells us more about the effect of these illusions than he does about how he created them, but the effects themselves must remind us of Fazio's description of van Eyck's landscape. Both are examples of illusion at what I have called cartographic scale.

This is what we read in the fragmentary *Vita anonyma*, now more or less agreed to be by Alberti himself:

> In the art of painting Alberti did unheard of things, incredible to those who saw them, which he showed in a little box with a small opening. There you might have seen huge mountains, vast provinces, the immense round of the surrounding sea, and regions so distant that the eye was dimmed. He called these things "demonstrations," and they were such that expert and layman alike would swear they saw, not painted things, but real ones in nature. There were two kinds of demonstrations, those he called diurnal and those he called nocturnal. In the nocturnal ones you see the North Star, the Pleiades, Orion, and that kind of bright signs.

To this point, Alberti's description suggests that a wide expanse of the sky was shown, about 150 degrees. He continues that there was also to be seen "the dawning moon rising from the lofty summit of cliffs and heights, while the evening stars were shining. In the diurnal demonstrations shone the star, that, to say it with Homer, shines after Aurora, mother of light, amply illuminating the immense terraqueous orb [that is, the sphere of earth and water]."[26] Alberti tells us that sailors (who would routinely have used portolan maps) admired his demonstrations, especially "certain Greek captains, to whom the sea was well known, for when he showed them this artificial immensity of world through, as I have said, a little opening, this is how they responded: 'Ah,' said they, 'we see a fleet of ships in the midst of the waves. It will be here by noon, unless that cloud in the East of the sun and that awful approaching storm stops them. We see, too, that the sea has grown wild, and a sign of danger is the too acute angle of the sun's rays reflected from the sea.'"[27] Since it was morning in Alberti's painting and the ships to be seen might be expected to reach port at midday, it may be surmised either that there were ships in the foreground and a port in the far distance, or—this is more likely—that ships were visible on the horizon, hours from the port in which the painting placed the viewer. Either way, the distances depicted must have been great, and they must also have been represented so clearly that experienced sailors could estimate them. Alberti's dawn sky must also have been red—red in the morning, sailor take warning, red at night, sailor's delight—the twinkle of waves indicating a sea stirred by winds from the direction of the rising sun. In any case, much as the horses whinnied at the painted horses of Apelles,[28] skilled seamen not only saw in Alberti's painted dawn an approaching fleet but felt confident to estimate its speed, distance, and direction and to warn of dangerous weather on the basis of painted sky and sparkling, agitated seas. We may think again of the expanse of the sea giving up its dead in van Eyck's Metropolitan *Last Judgment* (figure 3.9), the horizon with distant ships and the sweeping vista of the *bas-de-page* of the *Cavalcade by the Seashore* from the Turin-Milan Hours (figure 4.13), or the rough waters of the *St. Julian Ferrying Christ* from the same manuscript.[29]

We know that Alberti performed his miraculous demonstrations between 1432 and 1435 because he placed them in Rome and referred to them in the first, Latin version of his treatise on painting. Having outlined the geometry of vision in book 1, Alberti turned to the question of color.[30] Here the differences between the Tuscan and the terser but clearer Latin version are significant. Shadows occur when rays of heavenly or earthly light are "intercepted," but this interception also entails reflection. That is, when light is blocked by an opaque object, the rays are not simply stopped or absorbed, thus to cast a shadow; they are also reflected. Furthermore, Alberti said, reflected rays either return to their origin or go elsewhere. Alberti's final point in both versions is that reflected light carries the color of the surface from which it is reflected, as, he said, we may see in the pale green on the face of someone walking through a meadow in sunlight. On the way to this conclusion, however, he said in the Latin version that light rays obey the fundamental law of reflection, that the angle of incidence equals the angle of reflection, "as the mathematicians prove." But this, he interjected, "pertains to another part of painting." In the Tuscan version, the mathematicians are not mentioned, and we read instead, "One may say many things about these reflections, which pertain to those miracles of painting that many of my companions saw done by me at other times in Rome."

The second reference to his "miracles" appears as a pause in Alberti's "window" construction. He had just began to address the problem of placing transversals, observing that, "as no learned person would deny, no depicted thing may be seen to be equal to a true thing unless placed with certain reason."[31] He then presented his own solution to the placing of transversals—the viewer located at a point determined by modular size and distance from the side of the panel, which becomes the plane of intersection of lines of sight from viewer to baseline. Alberti left the theoretical explanation of this procedure for another day. "We will explain the reasons for this if we ever write about those demonstrations of painting, which our friends marveled at when we did them, and

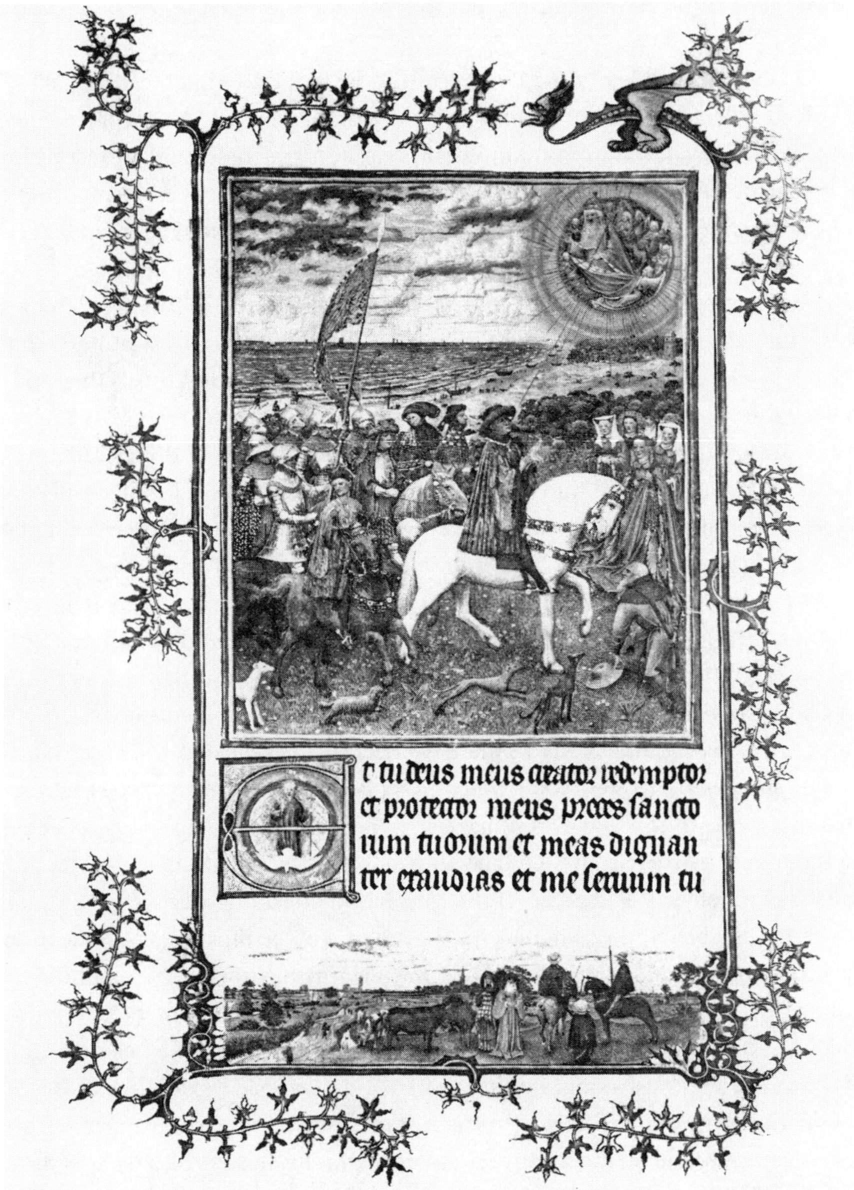

FIGURE 4.13. *Hubert and/or Jan van Eyck,* Cavalcade by the Seashore, *1415–17 or later, formerly Turin, Museo Civico. From* Heures de Turin, *pl. 30.*

called them 'miracles of painting.' For to that same part what I have said greatly pertains."[32]

What is to be made of this? It must at least be surmised that these "miracles" involved reflection and the *geometric principles* of reflection. And whatever device he may have used in his demonstrations, a mirror reflection—most probably a plane mirror reflection—was part of it and had to be seen from a single modular point, like the paintings Alberti was describing when his mind turned back to his Roman "miracles." In these demonstrations, viewing distance was enforced by a box, and the important common principle linking Alberti's miracles with the construction he described in his treatise on painting is that *the world we see is not so much a world of measures as of ratios*; this is the world reflected in the mirror of the eye. As we have seen, Alberti wrote that the dimensions of the world might be halved, and if the relations of things remained the same, no one would know the difference.[33] Alberti's Greek sailors, if they viewed his "miracles" under the right conditions, would see what seemed to be the vast expanse of heavens, land, and sea in a small box.

Any mirror reflection, true as it might be in the static abstract, is utterly fugitive and can only be seen from a single point in space and, for all practical purposes, in time. Perhaps Alberti used a mirror in the way Filarete described, to transpose visible forms into two dimensions as an aid to their imitation.[34] In this case, however, the forms would have been the very distant forms of landscape, again *on the assumption that an image in a plane mirror is a true image*. A painting made in this way would have shown that color is in all cases reflected in light; even if their colors change with intervening atmosphere, we see the most distant objects as colored. If the viewer looked through a hole in the back of the landscape painting (done with the aid of a plane mirror?) at a reflection in a plane mirror at the opposite end of the box, then Brunelleschi's (and finally Alhazen's) point about the communication of facing surfaces—of *all* facing surfaces—and the equivalence of eye and mirror would again have been demonstrated. The demonstration would in this case have been extended from the prismatic forms of architecture to the distant, indistinct, and unconstructable forms of landscape or, more properly, of cartography and astronomy. It is most important that painting, cartography, and astronomy are brought together by the same geometric representation. Although point of view was built into Alberti's miracles, point of view cannot have the same significance here that it had in the urban vistas of Brunelleschi or in Alberti's own later theatrical constructions. In these "miracles," human sight was placed once and for all in relation to the world at large.

The astronomical scope of Alberti's nocturnal painting suggests the scale of his miraculous illusions of land and sea. Alberti was a respected astronomer, and it has been argued that he designed and painted the astrological "night sky" in Brunelleschi's Old Sacristy.[35] (The earliest European sky maps, it will be remembered, were made in Vienna as a dimension of the cartographic studies I discussed in the last chapter.)[36] The disciplinary richness of Alberti's miracles is inversely proportional to the simplicity of the means. Alhazen's principle of reflected light, together with the ancient demonstration of the truth of images in a plane mirror, meant that planar reflections at whatever distance are true. This might suggest that Alberti (and perhaps Jan van Eyck) used plane mirrors as an aid to painting their landscapes, because, knowing planar reflections to be true in principle, they knew that the distant world was as susceptible to measure as the world at hand.

The theme of mirrors may now take us in a new direction, to one of the most legend-shrouded questions in the history of Renaissance art: the origin and diffusion of oil painting. Vasari attributed the invention of medium and technique to Jan van Eyck around 1410 and wrote that his secret was brought to Italy by Antonello da Messina.[37] There are earlier examples of the use of oil as a vehicle for pigments, but these cannot be regarded as equivalent to glazing and to the exploitation of the translucency of pigments in oil and varnish for the creation of a whole visual field. Given the fundamental difference between oil as a possible vehicle and glazing as an overall technique, it has also become clear that the process of making paintings in the Eyckian manner, rather than requiring a great number of fine glazes with mysterious concoctions, was fairly straightforward, although it had been brought to a new and utterly different level of technical, descriptive, and expressive possibility.[38] The question then is not simply the use of oil as a vehicle but rather what the reasons might have been for the development of these new and qualitatively different skills.

Masolino is the first Italian painter so far shown by modern analysis to have used oil as a medium.[39] He used oil exclusively as a vehicle for some pigments and sometimes combined it with traditional tempera—in what is called "tempera grassa"—in the Santa Maria Maggiore altarpiece he painted with Masaccio in 1427–28. (In this case, Masaccio was the more conservative of the two, sticking with pure tempera.) It has been surmised that Masolino learned this technique during his stay in Hungary, which happened to coincide with van Eyck's secret mission to distant places. Perhaps both found themselves in Hungary. In any case, no differences in style between the panels of Masaccio and Masolino for the Santa Maria Maggiore altarpiece were noted before modern analysis, and there is no reason to suppose that adaptation of a new vehicle to habits formed in years of training and practice, without the whole cluster of procedures of which glazing was part, would yield many differences. It is this difficulty of technical transfer that brings us back to the subject of mirrors, and I am going to argue that the mirror provided the metaphor, the imagined principle of unity, for Jan van Eyck's painting.

The words "mirror without stain" are taken from what Millard Meiss called Jan van Eyck's favorite biblical text, to be seen on the border of the Virgin's tunic in the Berlin *Madonna in the Cathedral*, in the panel of the Virgin in the Ghent Altarpiece, and on the frame of the *Madonna of the Canon van der Paele*.[40] The text is from the *Wisdom of Solomon* 7:29 and reads in part, "She is more beautiful than the sun above all the stars," and she is "the mirror without stain of the eternal light of the majesty of God." The identification of the Virgin Mary with glass and of light with the Holy Spirit was not limited to repetition of this one text. The lost frame of the Berlin Madonna was inscribed with the words of a popular hymn, "As the sunbeam through the glass passes but not stains, so the Virgin, as she was, a virgin still remains."[41] In chapter 2 I compared such metaphors to Alhazen's description of the crystalline humor.[42]

A mirror is a sheet of glass backed with lead or silver, and in what follows I will urge the conclusion that an Eyckian painting is itself analogous to a mirror. Heinrich Schwarz, in a study of the use of mirrors by painters in the late Middle Ages and the Renaissance, noted that painters in Bruges belonged to the same guild as mirror makers.[43] In Florence, as Roberta Olson pointed out, both painters and mirror makers belonged to the guild of Medici e Speziali, presumably because both bought materials from pharmacists.[44] As I remarked in passing in chapter 2,[45] Brunelleschi would thus have had expert advice close at hand on matters relating to mirrors, and perhaps in Flanders this guild collegiality had other consequences. Whereas Brunelleschi was primarily interested in the geometry of the plane mirror, like concerns in the North yielded another variant: an Eyckian painting is like a mirror in being a continuous reflective surface over which a continuous transparent surface is laid.

As we have seen at several points, it was a principle of ancient and medieval optics that the eye itself sees only light and color. Then, by means of faculties other than the organ of vision itself, we make other judgments—for example, of distance.[46] In these terms, a panel with a white ground is a reflective surface, but it is also light on which any visible mark becomes dark by contrast. Underpainting thus becomes the organization of the image in basic terms of value. With this in mind, we might hazard a guess as to what authors Fazio's "not unlettered" Jan van Eyck read other than Pliny the Elder. As I discussed in the introduction, Aristotle and his many Scholastic commentators located all color between the extremes of white and black, so glazed pigments might be seen both to preserve and to differentiate the first framework of value contrasts, into which colors are set.

We may consider van Eyck's little *St. Barbara* in Antwerp (figure 4.14). A faux red marble frame was always attached to the wood panel, and on this frame the painting is signed and dated 1437. Van Eyck, rather than showing the saint holding her tower, shows her seated on a little plateau in a posture of humility, daintily reading, holding her long palm of martyrdom, while the elaborate late Gothic structure in which she would be housed rises in the background. At a distance behind her to her left, ladies approach, and to her right workmen of all kinds busy themselves. Inside the shadowed arches above her head the great wheel for hoisting stones may be seen. The landscape with a distant hill town reaches to the horizon.

Some writers consider the *St. Barbara* unfinished, in which case it is an underpainting; others argue that it is too carefully elaborated to be an underpainting and that it is finished as it stands; still others, completing the possible scholarly permutations, argue that it began as underpainting but then was worked up as an unusually fine drawing. Whatever the case, the image is realized and unified entirely in terms of value contrast and modulation, and in this way it unifies the whole image and its surrounding space as a continuous field of fundamental and organizing visual elements. A diffused light falls from our upper left, casting a soft shadow beneath Barbara's left elbow. Contours are not lines but rather always edges of light against shadow, and the hems of her garments, rather than being defined by lines, cast thin shadows just like those in the doorway above her right shoulder.

The wood panel was prepared with layers of chalk and glue, and the paint, applied with the tip of a fine brush, is black, modeled through innumerable grays by thinning and hatching. A certain amount of drawing by incision and scraping with a stylus is to be seen. There is debate about the highly tinted ultramarine sky, which most seem to prefer to think was painted later. Touches of lead white oil paint, however, are to be found throughout the painting—some may be seen between the windows of the tower—so that there is a full range of values between black and reflective white.[47] (It might be suggested that van Eyck noted the brightest parts of the painting first so that shadows and colors could be adjusted to them. In that case, the painting would be unfinished.)

Whether or not the *St. Barbara* is finished, the painting displays the tonal understructure of a painting by Jan van Eyck, which is simple and disciplined but at the same time as endlessly capacious as vision itself—or as mirrors themselves. Had the *St. Barbara* been colored—like the *Madonna and Child (Lucca Madonna)* in Frankfurt (figure 4.15)—the reflectivity of the white ground would

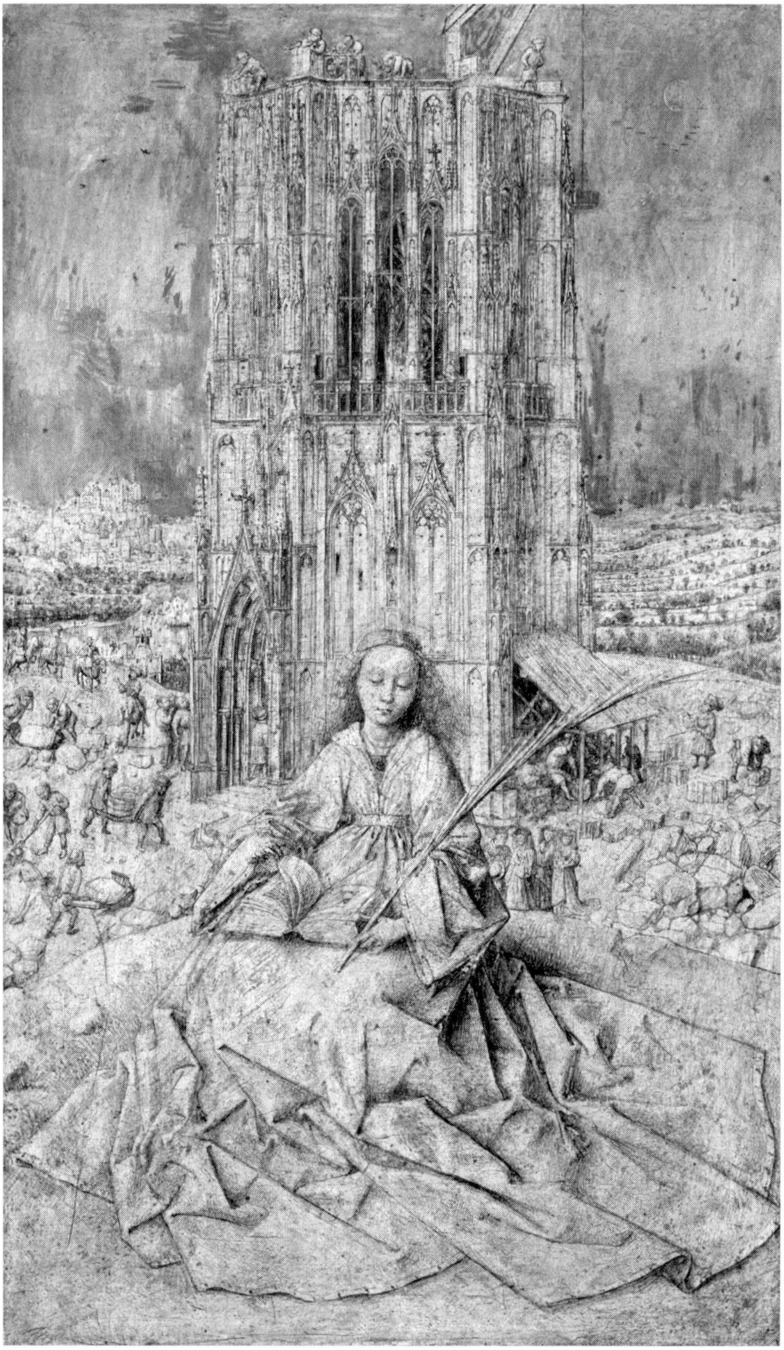

FIGURE 4.14. *Jan van Eyck,* St. Barbara, *underpainting on panel, 1437, Koninklijk Museum voor Schone Kunsten, Antwerp. Scala/Art Resource, New York.*

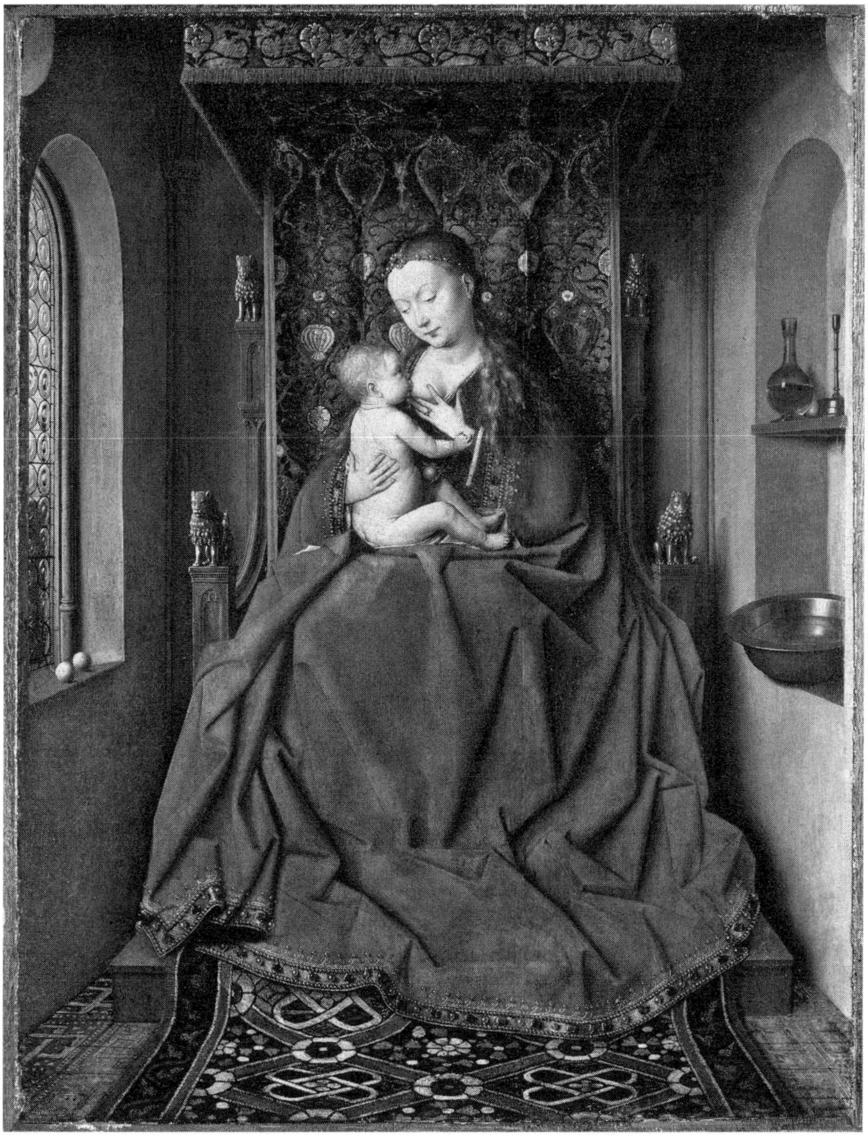

FIGURE 4.15. *Jan van Eyck,* Madonna and Child (Lucca Madonna), *oil on panel, ca. 1434–35, Städelsches Kunstinstitut, Frankfurt. Photo © Ursula Edelmann, Artothek. (Reproduced in color as plate 11.)*

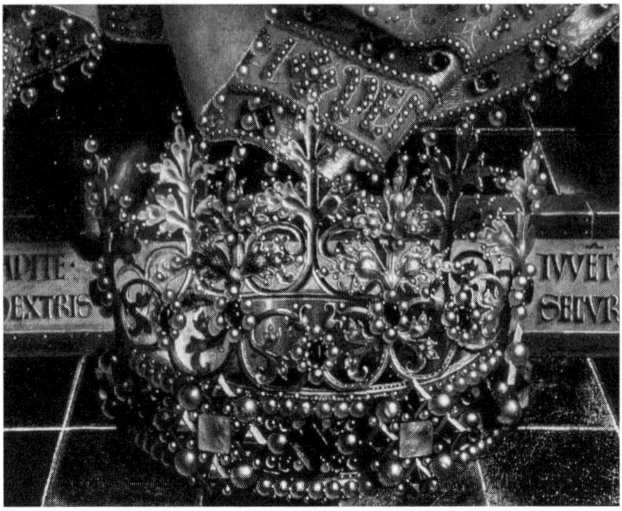

FIGURE 4.16. *Jan van Eyck, Ghent Altarpiece, open, detail, crown at the feet of God the Father, completed 1432, Cathedral of St. Bavo, Ghent. Scala/Art Resource, New York.*

have been exploited by glazing, shadows would have been deepened, and lead white highlights, opaque and themselves reflective, would have been selectively and systematically added. At this point, Jan van Eyck's painting should remind us of Alberti, who told us that a mirror is the best guide for the disposition of light and shadow in a painting and that white should be sold to painters at the price of precious stones so that it would be used with the utmost economy.[48] He insisted that "the painter will find nothing but white to represent the brightest glow of light," that "no surface should be made so white that you cannot make it a great deal whiter still."[49] Perhaps Alberti had already learned this from Eyckian painting if not from van Eyck himself.

Technical analysis of the Ghent Altarpiece revealed a change, a major pentimento, right at the symbolic heart of the image.[50] A crown was inscribed in the gesso of the panel at the feet of God the Father (figure 4.16), together with an inscription, but only the inscription was painted on the tile floor. This first version was scraped down and covered with silver foil, over which crown and floor were painted as we see them. Some have concluded that this repainting was done in the sixteenth century, but if Jan van Eyck made the change himself, then the procedure he followed is even more explicitly like the making of a silver-backed mirror. In one case, reflective metal is applied to the back of a glass surface; in the other, pigments in translucent oil are glazed over a reflective surface, either white or silver.

If we think of Eyckian paintings of the Virgin as like the mirrors to which they allude symbolically, they are unusual and special mirrors in which images, although like objects of our sight, have the independence of their own right and

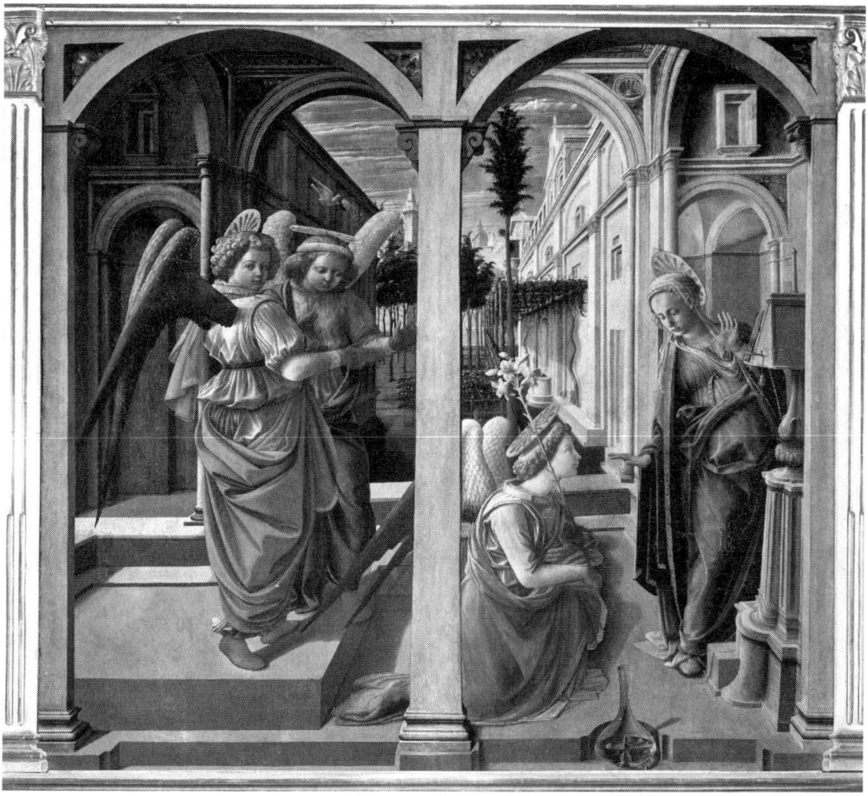

FIGURE 4.17. *Fra Filippo Lippi,* Annunciation, *panel, ca. 1440, San Lorenzo, Martelli Chapel, Florence. Scala/Art Resource, New York.*

left, just as planar order and supernatural, hierarchical relative scale remained important parts of van Eyck's visions. These paintings are also mirrors in which we do not see ourselves—or, more exactly, we do not see our mundane, physical, and optical selves. Literally in place of ourselves we may see purity, deity, spiritual example, objects of devotion and eschatological longing. (The fact that one of the paintings Fazio described showed women of uncommon beauty stepping from the bath is another story, although not necessarily an altogether incompatible story.)

It is not surprising that the new domestic format of the Florentine tondo, which mimicked the round convex mirror or could even be seen to be like such mirrors, treated similar themes.[51] Filippo Lippi, who painted some of the earliest tondi, was also especially interested in—or catered to patrons who were especially interested in—the religious symbolism of light. We might consider the water flask in the Martelli *Annunciation* in San Lorenzo (figure 4.17)—to be

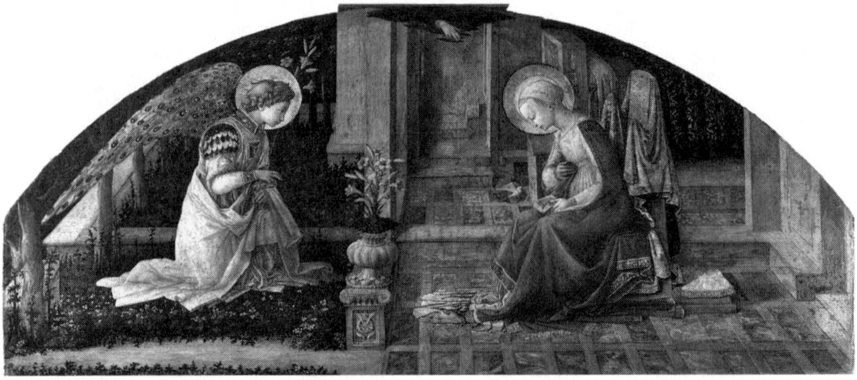

FIGURE 4.18. *Fra Filippo Lippi,* Annunciation, *panel, 1449, The National Gallery, London. Photo © The National Gallery, London.*

compared to the flask in the niche to our right in figure 4.15 or to that in figure 4.7—or the Medici *Annunciation* (figure 4.18).[52] These are not tondi, but in one of the earliest examples of the form, the *Virgin and Child* in the Pitti (figure 4.19), painted in the early 1450s, the mirror is doubly, deeply immaculate. In the background to our right we see the meeting of Joachim and Anna, and to our nearer left, the lying-in of Anna, who has just given birth to the Virgin. There are thus two immaculate conceptions, one of the Virgin, the other of her son. As in Florentine *deschi da parto*, the Florentine birth platters to which the origin of tondi has also been traced, servants attend and women bring gifts. The sacred continuous narrative takes place in separate perspective compartments but also in what is clearly the interior of a Florentine palazzo. These references lend the sacred narrative a deep familiarity, but the owners could not have seen themselves in these everyday surroundings, or, if they did, they saw themselves reflected as spiritual exemplars of purity, humility, and virtue.

There is a long and broad tradition, classical and Christian, of "exemplary mirroring"[53]—or, perhaps more precisely, of the moral significance of mirror images. Seneca dismissed rainbows as no more than reflections, "deceptions of mirrors," leading to his thoughts on mirrors in general. "What is shown in a mirror is not there,"[54] and mirrors may distort and magnify beyond all recognition. He told the story of a libertine who surrounded himself with mirrors in order to watch the polymorphous orgies in which he indulged, using concave mirrors to enlarge details to dreamed-of dimensions. Warped mirrors reflect warped behavior, an extreme case of the lust and extravagance of which mirrors had become part in his own time. Seneca provides a short history. Humans first saw their own faces by accident, in "clear fountains or polished stones."

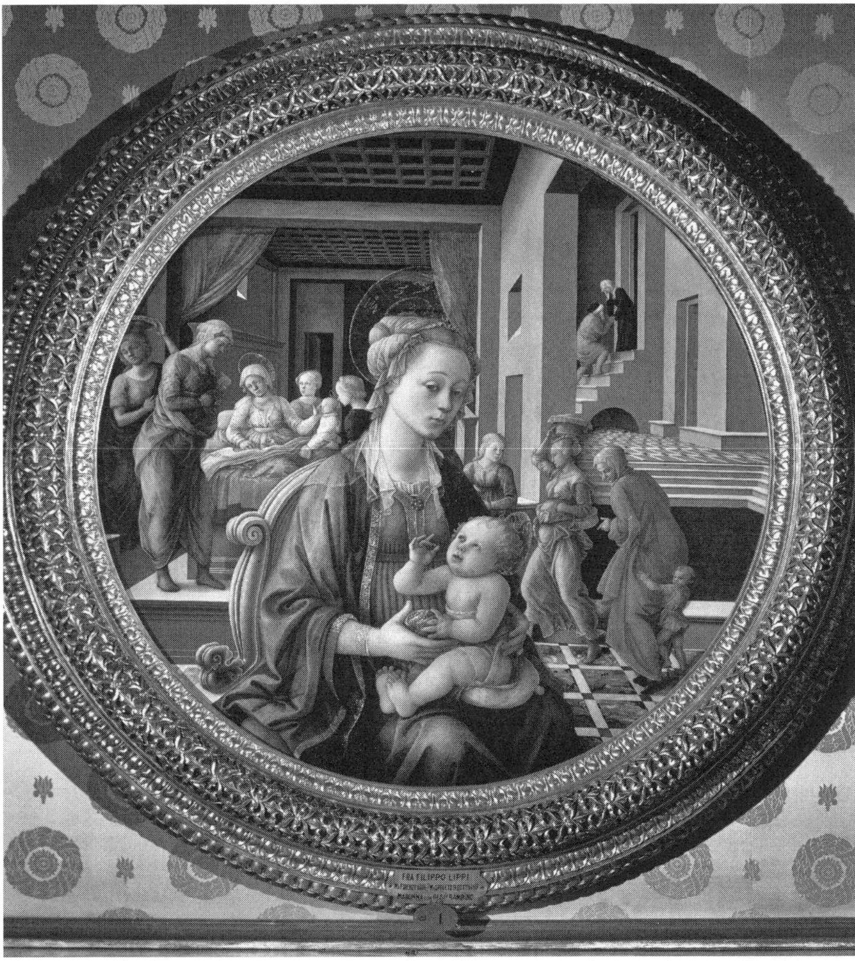

FIGURE 4.19. *Fra Filippo Lippi,* Virgin and Child with the Meeting of Joachim and Anna and Birth of the Virgin, *panel, 1452, Galleria Palatina, Florence. Scala/Art Resource, New York.*

"Then, when love of self, innate in mortals, had made the sight of their form pleasing, men looked down oftener into those surfaces where they saw their own images." The decadence had begun. Mirrors, however, offer a choice, and they were also invented "in order that man may know himself,"[55] thus to fulfill the Delphic and Socratic injunction. (We may still say, "Take a good look at yourself!") The homely man, Seneca said, learns that he must compensate with virtue; the old man learns to give up behavior inappropriate to gray hairs and to give some thought to death. The principle of exemplary mirroring could be extended to "reflections" of ethical and religious norms and ideals in general, from the "mirrors" of princes as merciful and just to the *Madonnas* and *Holy Families* of Florentine palazzi.

Alberti began the second book of his treatise *On Painting* with an exhortation to youths inclined to undertake the arduous path to mastery of the art, which, he said, possesses a "divine force." Like friendship, painting makes the absent present and makes the dead seem alive. Gold is more valuable when worked by a great artist, and lead formed by Phidias or Praxitetes is more valuable than unworked silver. Alberti followed Pliny in reporting that Zeuxis gave his works away because no price could be sufficient for them but embellished his source when he explained why. How could one be satisfied who, "in modeling and painting animated things, might show himself to be almost another god?" Alberti repeated himself: any "master painter will see his works adored, and feel himself judged almost another god."[56]

According to Alberti, painting gives *decorum* to everything, by which I take him to mean that painting embellishes everything in appropriate and pleasing ways. In the Tuscan version, stress falls on "*bellezza*" rather than "decorum": "wherever you find beauty in things, that may be said to be born from painting."[57] These remarks about decorum and beauty immediately precede those in which I am most interested, the attribution of the invention of painting to Narcissus. "Therefore I used to say to my friends, following the thought of the poets, that Narcissus, who was changed into a flower, was the inventor of painting; because, [painting] being the flower of every art, this myth [*fabula*] is entirely appropriate [in the Latin, beautifully appropriate]. Who would tell you that painting is anything else than similarly to embrace [*abbracciare*] with art that [which is] there on the surface of the pool?" (The Latin *amplecti* perhaps has stronger connotations of ardor.)[58]

It is important that Alberti, who claimed to have no precursor for his new art, called Narcissus the "inventor" of painting. Alberti reviewed the ancient myth of origin, citing not the story from Pliny we considered in chapter 1 but Quintilian, who "believed that the first painters began the art by circumscribing shadows of the sun."[59] It did not matter to Alberti if the Egyptians invented the art, he said, because he was not interested in writing a history like Pliny, and in any case, the writings of the ancients (as well as the paintings) did not survive. Rather, Alberti said, "we are fashioning anew [*di nuovo fabrichiamo*] an art of painting" (in Latin, "we are recounting an altogether new art [*artem novissime recenseamus*]").[60] The paradigm of this new art of painting is not the shadow but rather reflection—and *self*-reflection.[61]

Alberti may have rejected the ancient historians, but he also appealed to the poets. Cecil Grayson noted that no ancient poet called Narcissus the inventor of painting,[62] and Alberti may simply have misremembered his sources. But the *ekphrasis* on a painting of Narcissus in the *Imagines* of Philostratus the Elder begins as follows: "The pool paints Narcissus, and the painting [Philostratus is describing] represents both the pool and the whole story of Narcissus." We may be fooled by the painting, but Narcissus is fooled by the artifice (*sophisma*) of the pool; he is so engrossed by the "painting" of his own image that he cannot hear us, and so we must interpret the painting of him and his reflection for ourselves.[63] Alberti read Greek,[64] so the Philostrati, Elder and Younger, were available to him, just as they must have been of interest as evocations of lost ancient painting. There is also another connection. Baxandall argued that the prooemium to the *Imagines* of Philostratus the Younger was the model for the introduction to Fazio's praise of painters in his *De viris illustribus*.[65] Like Fazio, both Philostratus the Elder and Philostratus the Younger likened painting to poetry, and so to praise one art was to praise the other.

In order to be changed into a flower, it was necessary for Narcissus to fall in love with his own reflection as if it were another, and then, when he had realized that it was just like himself but without body, to pine away because, like Echo, the nymph whose love he spurned, he could never possess the beauty he could only see. The image in the mirror, as we have seen, is true but is not real. It is difficult to read Alberti's words without remembering his autobiography, written in the third person, which closely parallels the arguments I have just reviewed. Alberti wrote of himself that

> he used to invite his friends and have with them endless discussions of literature and philosophy; and to these friends he would also dictate little works while he painted the portrait or made a wax model of one of them. In Venice he modeled the faces of his friends in Florence whom he had not seen for a year and whole months. He used to ask some young boy whether he recognized the image he was painting, and if it was not recognizable to a child, he denied that the painting was a work of true art. He strove to render his own

features and characteristic appearance, so that, by the painted or modeled image, he might be readily known to strangers who summoned him.[66]

Alberti wrote that he was always examining his own conduct so that no one might suspect anything ill of him. Slanderers, however, can do irreparable damage even to the best-guarded reputations, and to counteract slander it is necessary for "everything in life, every gesture and word, both to be *and to appear to be* [*et esse et videre*] worthy of the good will of good people." This demands art, doubly so because this art must be made to seem artless. Alberti has invoked two ancient and fundamental rhetorical analogies, ornament and the "art of concealing art." Both have become social virtues. Something like an eloquence of action, ornamented and therefore pleasing to all, should be evident in how one walks in the street, how one rides, and how one speaks. But the artifice of this ornament must never be obvious. If it is, then the *appearance* is no longer persuasive, and one falls into affectation and vice. The passage from stylistic to ethical is crucial. In ancient rhetoric, the solution to the problem raised by eloquence—that it may persuade of the false—was the formula *vir bonus dicendi peritus* (the good man skilled in speaking).[67] Alberti certainly knew this formula, but in his version of it there is a fundamental change in emphasis. Individual goodness is evident through and in a more general eloquence, hidden by a higher artifice, the surface of which might be shattered at any time. This artful appearance is always addressed to an audience of the good, which is, however, a slippery standard, especially when dealing with those of a similar eloquence. It is most fundamental that the success of the enterprise is evident in the esteem of others—or, one might say, it is *mirrored* in the esteem of others. In such circumstances, the "good" might amount to no more than pressure to conform to certain norms or to curry favor, but it might also provide the occasion for the creation of the self one would like to be, so that the sickly and penurious Battista became Leo, the noble lion, indifferent to the dogs nipping at his heels, skilled horseman and athlete, the first universal man.

If, as these last remarks suggest, there are several ways in which Alberti might be said to be Narcissus, we should also not forget that Alberti occupies an important place in the history of portraiture and of *self*-portraiture. Vasari reported having seen a self-portrait by Alberti, a painting made with a mirror.[68] Alberti's bronze relief self-portrait (figure I.1), showing him *all'antica* (in profile), must have been made with *two* mirrors. It occupies an early place in the history of Italian Renaissance portraiture taken altogether and is shaped by the spirit of his device, by his physiognomy, as he renames himself Leo and shows himself with the scruffy hair and clouded brow of the leonine temperament.[69]

Alberti's analogy to Narcissus breaks down into two parts. Narcissus, he seems at first to say, was the inventor of painting for no other reason than that he turned into a flower and painting is the flower of all the other arts, giving them whatever decorum and beauty they possess. But then Alberti said that the story of Narcissus is itself "beautifully appropriate"—that is, the story of Narcissus itself makes the origin of painting evident in a suitably embellished way, a figured, poetic way. Furthermore, painting is the art that itself achieves such embellishment. What does painting embellish? It embellishes—or, as Alberti puts it, "embraces with art"—what is seen "there on the surface of the pool." The primary reality is in fact the points on the surface of the mirror, points that may be marked, as Toscanelli argued, but then are gone, as Toscanelli did not argue, leaving nothing but scant traces of time past.

The myth of Echo and Narcissus is about reflection, the turning back of sound, the turning back of light. The pool upon which Ovid's Narcissus chanced was never warmed by the sun, its water was clear and silvery bright, and it was his alone.[70] No one else ever came near the pool; nothing ever disturbed it. The pool was and always had been, in short, a perfect plane mirror. In it or on it, Ovid said, "he loves hope without body, thinking body to be what is shadow." What separated Narcissus and his beloved? It was, Narcissus himself said, mere water: there was no ocean that might be crossed or mountain that might be climbed, no real obstacle to be overcome to attain his love. This takes us back to Alberti's rhetorical question: "For what else is it to paint than to embrace through art that surface of the pool?" Ovid did not refer to the surface of the pool, but Alberti wrote of *superficies* in Latin and *superficie* in Italian. For Alberti, these are fundamental analytic terms, used throughout the first book of his treatise on painting to refer to the colored surfaces of visible forms; here it refers to that reality framed, the whole painting *as and for* vision. It is Alberti's transparent window, a synoptic and subjective surface, the unifying surface of the surfaces of things in light, all making up a seamless, framed field of light and dark, not so much a *point* of view as a *surface* of view, more or less equivalent to a living mirror.

The myth of Narcissus had been moralized as an example of vanity long before Alberti, and in his commentary on Plato's *Symposium*, completed in 1469—while Alberti was still alive—but published in 1484, the Florentine Neoplatonist Marsilio Ficino, appealing to the authority of Orpheus and Plotinus, used the myth of Narcissus to what he must have intended to be crushing metaphysical effect.[71] His meaning is very different from Alberti's. Beauty, Ficino said, is nothing but light, which for us, however, is "stained" by multiplicity and spatiotemporal particularity. We may think these stains away to contemplate pure light, but we must also know that as the world descends from pure light, each stage in this descent loves the next lower stage but does not forget to love itself. Only the human soul, which is free to do so, differs in this regard, forgetting its own beauty—closer to the light—while transfixed by the beauty of the body. This, Ficino says, is the plight of Narcissus, "racked by terrible passions and stained by the filth of the body." The soul dies, as it were, "since it now seems to be a body rather than a soul." We have not encountered in Alberti's arguments the least trace of such emanationism, which placed all matter lower than all light. For Alberti, light was physical, as it was in the optical tradition on which he drew. The reflected image of the body was physical and, as physical, was an irreducible reality. Particularity could not be a "stain."

What does it mean to embrace the reality of this surface? This is a big question, not least because the idea of Narcissus has expanded to truly world-historical proportions since Alberti chose this myth to allegorize the new art of painting. Sigmund Freud defined the concept of narcissism around 1910, when he wrote and published his study of Alberti's younger contemporary and fellow love child, Leonardo da Vinci, and from these psychoanalytic beginnings narcissism has expanded to become part of the functional language of Western modernism. The contradictory strains in Alberti's writing have been explained in terms of narcissism,[72] of a personality based on swings between the absolutely regressive poles of infant omnipotence and frustration. However true that may be, Alberti's appeal to Narcissus came in the context of unprecedented concern with the optical and with the world to be reimagined in terms of a universal opticality. It was in this context that Alberti's invocation of Narcissus was an awakening that assumed prophetic depth. Alberti could hardly have foreseen narcissism as a condition of Western modernity, as the psychological basis for effective strategies in the vast institutions of consumer societies, or as the diagnostic basis for the criticism of these same institutions. Still, whether or not we are psychoanalysts, we have a pretty good idea of what is meant, for example, when someone says we all live in a "culture of narcissism,"[73] and Alberti—and the new painting of point of view and personal expression he helped initiate—may be legitimately placed near the beginnings of this new historical reality.

Even if Alberti's allusion, probably to the always-popular Ovid, is one of a long series of such allusions in art and literature, it would be a mistake to think that it was no different from earlier references and further to suppose that his version of the ancient myth is unrelated to the modern versions, however differently conceptually grounded they may be. Alberti's self-consciously modern retelling of the story of Narcissus in explicit relation to painting appears in the context of other, reinforcing themes of emerging modernity, and his Narcissus may be understood as a humanist version of the Florentine adage "every painter paints himself," which, as far as I know, was first attributed to Cosimo de' Medici.[74] *Maniera* could mean "individual hand" but also "individual vision"—that is, *fantasia*—and in both senses was essential to the animation of the image, the necessary ingredient by which a painting could be made to seem to live, given the spark of apparent life.[75] *Maniera* also only became visible on the surface, in the drawing or painting. It is when we do embrace the relative reality of point of view, the points marked by bits of wax on the surface of the mirror, the world *per comparazione*, which Alberti announced and praised as a prelude to the description of his perspective construction, that painting flowers, that beauty literally comes into view, that there are both apparent life and communication, and the painter is *not* another god but *almost* another god.

There are two traditions of the significance of light in the arguments I have made in this final chapter. There is the metaphorical optics of the "mirror without stain," a beautiful episode in the history of Mariology to be traced in both Northern and Southern Renaissance painting. This symbolic interpretation of light continues the tradition that reconciled the new optics to Christian theology. I have argued, however, that Brunelleschi and Alberti should be related to the more materialist Alhazen revival discussed in chapter 2. And if northern painting arose along different lines according to which reflection is more fundamental than the geometry of reflection, reflection in all cases absolutely locates or implies the location of a viewer. Both reflection and the geometry of reflection establish (or imply) point of view, and in this respect the two traditions are both close and early modern.

We may consider an example of the poetry of such location, such *optical* location, in Hans Memling's *Last Judgment Altarpiece* (figure 4.20), finished by 1467. Painted for Angelo Tani, an agent of the Medici bank in Bruges, this triptych was supposed to have come to Florence but wound up through the fortunes of war in Gdansk via London.[76] When Rogier van der Weyden painted his *Last Judgment* in the late 1440s, he showed Christ in Glory with a polished orb of the world at his feet, with a jeweled OT pattern (figure 4.21). To Christ's right on this orb is a triple reflection, perhaps of the lights of Paradise. In Memling's retelling, the orb at Christ's feet is a great spherical mirror; it reflects Mary to his right and St. John the Baptist to his left, and the universal judgment over which he presides, including the back of the winged St. Michael (figure 4.22). But it is most significant that in Memling's painting the world is no longer the ancient diagram of the *oikumene*, it is an all-reflecting and all-beholding mirror. Nicholas Cusanus must once again come to mind: in his *De visione dei*, he

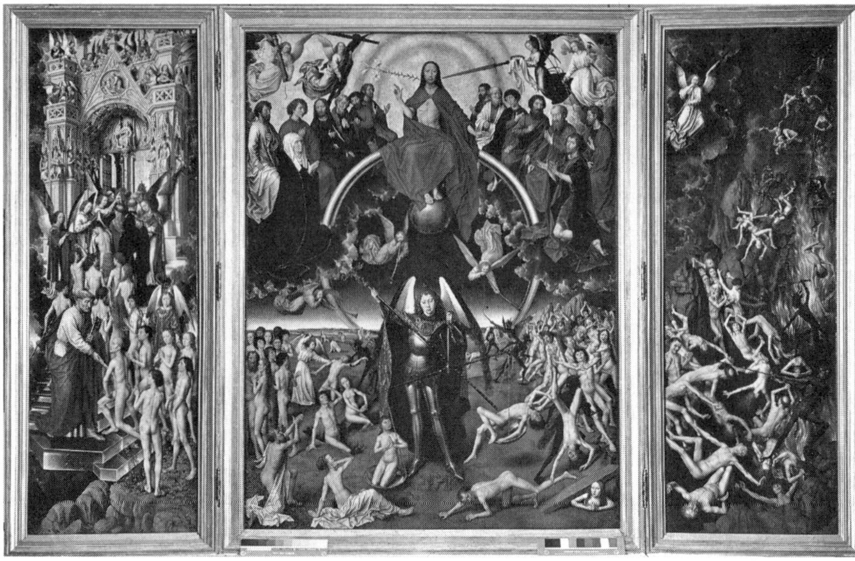

FIGURE 4.20. *Hans Memling,* Last Judgment Altarpiece, *open, ca. 1467, Narodowe Museum, Gdansk. Erich Lessing/Art Resource, New York.*

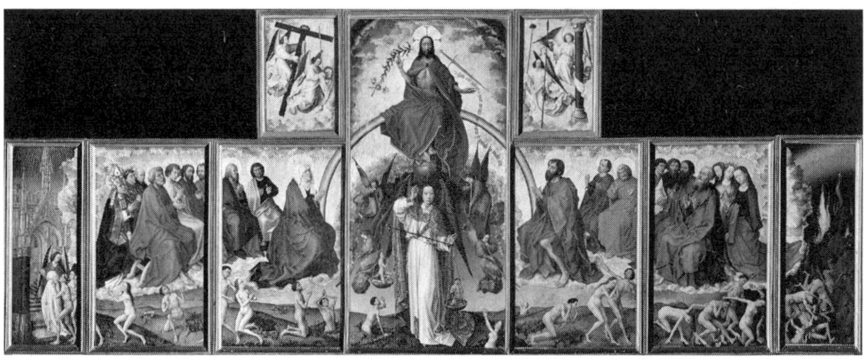

FIGURE 4.21. *Rogier van der Weyden,* Last Judgment Altarpiece, *open, ca. 1444–48, Hotel-Dieu, Beaune. Erich Lessing/Art Resource, New York.*

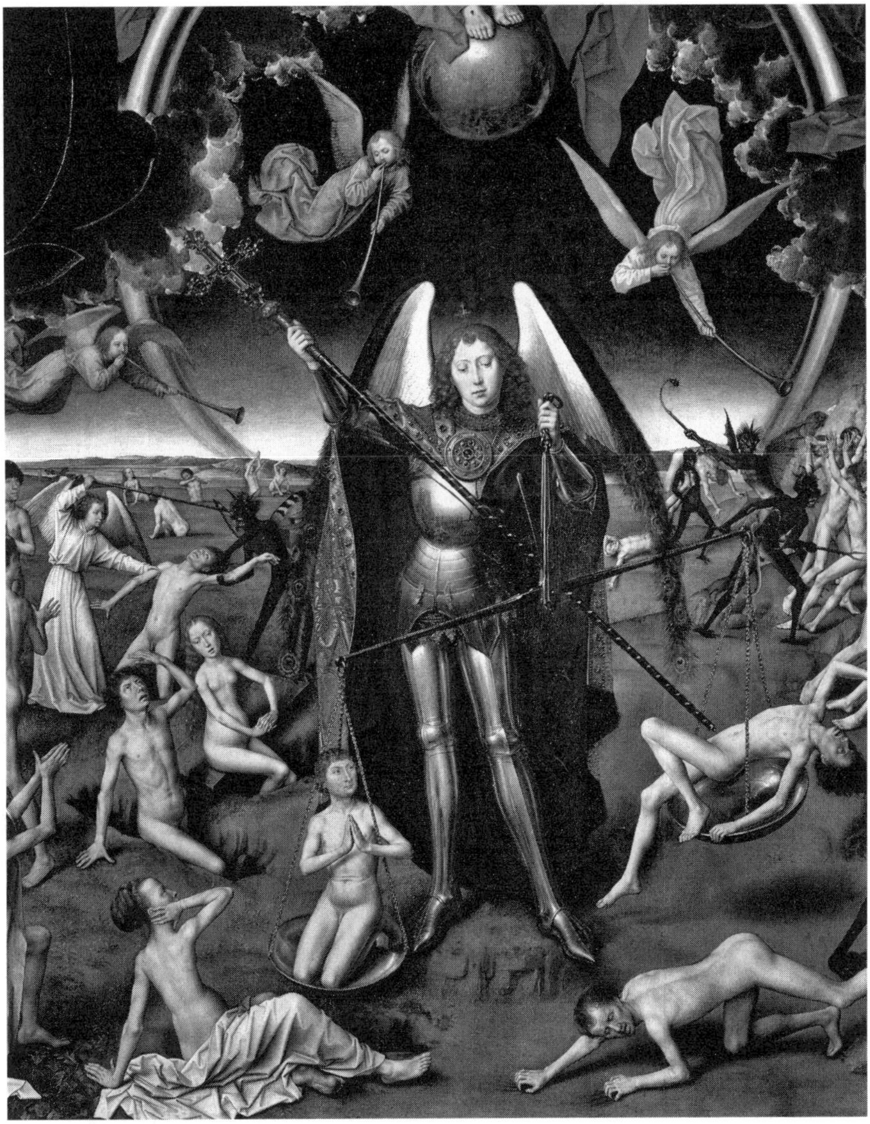

FIGURE 4.22. *Hans Memling,* Last Judgment, *detail of figure 4.20. (Reproduced in color as plate 12.)*

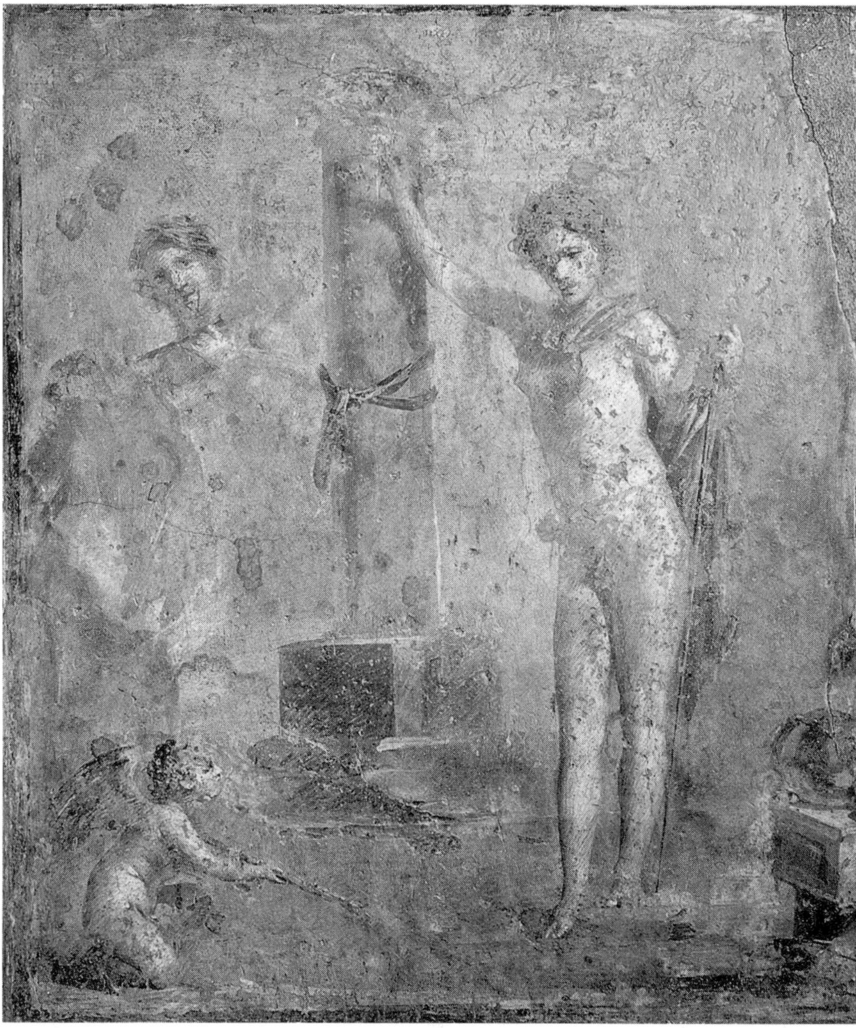

FIGURE 4.23. Narcissus, *Roman fresco, Pompeii, first century* C.E., *Museo Archeologico Nazionale, Naples.*

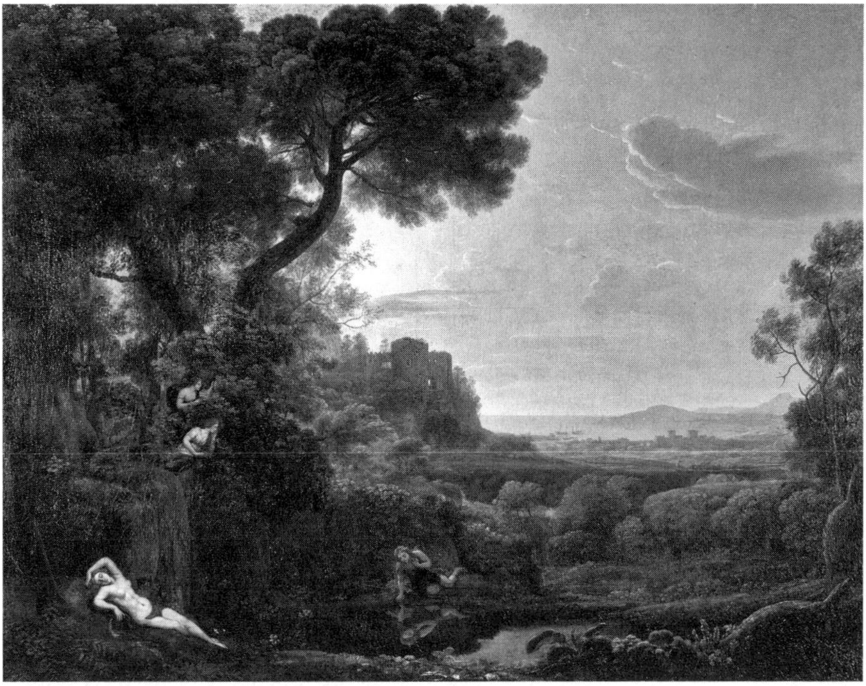

FIGURE 4.24. *Claude Lorraine,* Landscape with Narcissus and Echo, *oil on canvas, 1644, The National Gallery, London. Photo © The National Gallery, London.*

wrote, "Lord, you see and you have eyes. You are an eye, since with you having is being. You thus observe all things in yourself. If in me, my seeing were an eye as it is in you, then I should see all things in myself, since the eye is like a mirror. And a mirror, however small it be, beholds in itself the image of a great mountain and of all that exists on the surface of that mountain and so the species of all things are contained in the mirror of the eye."[77]

In the breastplate of the armor of St. Michael in figure 4.22, we see a reflection of the dead rising as if in the space before and around him. The nearest figures, including one saved (Tommaso Portinari?) in the pan of the scales Michael holds in his left hand, are carefully reversed; the blessed rise to Michael's right, and the last day stretches to an amber horizon, a visionary transformation of the horizon of space in which we stand. There are a number of contemporary portraits among the naked Paradise-bound, and it is possibly absolutely significant that because reflection inevitably implies location as surely as perspective construction does, the viewer might also look for a place among the rising dead.[78]

The theme of Narcissus is hard to represent for reasons having everything to do with the most basic optical principles. It is impossible for us to see Narcissus as he would see himself, and the reflection is—significantly enough in itself—usually left to the viewer's imagination. By the same token, it is impossible for us to look into a pool or a mirror without finding ourselves in the solitary predicament of Narcissus. In a Roman painting (figure 4.23), the gracile, self-absorbed youth is available for our enjoyment, but Narcissus would see—or imagine—himself very differently. Or perhaps we are meant to see Narcissus as the painter imagined Narcissus to imagine himself. However that may be, we see the form of the beautiful youth whose reflection Narcissus loved but not the reflection itself. Only a personified Eros, paradoxically touching fire to water, and a pining, wasting Echo tell of his futile self-desire.

The specular dilemma of Narcissus is perhaps best represented from a distance, as in the painting by Claude Lorraine in figure 4.24. The reclining figure to our left was repainted as a nude, changing the equation of the scene for the viewer. Whether or not this figure is (or was) Echo, Narcissus is absorbed in his own reflection. We see something like the "objective" Euclidean symmetry of form and reflection, but this is only an approximation from afar. Furthermore, distance in effect interprets the story for us as the image assumes the meaning of one seeing himself and therefore of our impossibly seeing ourselves seeing ourselves. Much as the paintings of Apollodorus of Athens "held the eye" because they presented sight to itself, Narcissus, the living mirror, sees himself (and we see ourselves) as the living subject. He is the unique subject of his own experience whose mortality consists of being constantly reborn to interaction with the world. When Cusanus wrote, "If I were to see as I am seen I should not be a creature,"[79] he meant that we are seen by God and cannot have God's vision. But we are also seen by others, and we cannot see ourselves as they see us. Nor can we live without seeing others and being seen by them.

Afterword

Perspective, Optics, and the Material Subject

From its beginnings, the Greek science of optics, like harmonics and astronomy, was a subalternate geometry: its "points" and "lines" were physical, and it used the arguments of its parent discipline as they were relevant. Of course light travels in straight lines, viewers may be said to stand at a "point," and Plato liked to pun on the shared sphericity of eye and sun as the kinship of the visible and the intelligible, a microcosmic-macrocosmic analogy that would shape the conception of the physiology of the eye to modern times.[1] For its part, the geometry in which optics nested was the first highly developed branch of mathematics, with many practical applications, from cosmology to architecture and surveying, all compatible with optics. The geometry of vision, however, was only the historical beginning of optics, and modern optics, nested in its turn in modern mathematics and physics, is concerned with the extrageometric aspects of the behavior of light. It is a physics of light and color or an equally specialized physiology and psychology of perception. As different as its emphases may be, however, optics in its modern forms is continuous with its ancient physical origin.

We are perhaps more inclined to think of sound than of light as physical. As I have just said, however, optics and harmonics were sister sciences, and Vitruvius, in addition to providing the fullest account of *skenographia*, described the theaters in which such illusions were to be seen. They were to be constructed in relation to the human voice, which, Vitruvius wrote, radiates in innumerable spherical waves from a central origin, much as, in two dimensions, waves spread from a stone dropped in water. If the architect does not understand this active geometry of sound, then waves in the theater will rebound and interfere with one another, muffling the actors' voices, and it was by the arrangement of theaters in accordance with the science of harmony that "the ancients increased the power of the voice."[2] Light is similarly physical, with its own geometry, knowledge of which contributes not only to the creation of illusion but also to the use and beauty of buildings.

As the offshoot of a subalternate geometry, the "points" and "lines" of Italian Renaissance painter's perspective were more material than the rays of light they represented, and useful abbreviation of an adaptation of optical principles was closely related to the art of drawing, which grew up in the closest conjunction in just the same years. The ongoing integration of painters' perspective—the schematic geometry of light—with the invention and composition of figures and with the composition and adjustment of light, dark, and color is fully as evident in Italian Renaissance drawing and painting as perspective itself. This also takes us to the extrageometric aspects of optics and leads us at the same time to the importance of imagination and the judgment of the eye.[3]

We spin very different worlds if we imagine human conscious presence in the world beginning from a point than we do if we imagine it beginning from a surface—in this case, the surface of the eye. Modern and contemporary discussions of perspective (often as representative of vision in general), usually proceed from the first alternative, relating it to notional Cartesian and Newtonian coordinate geometry. In these terms, vision and its mechanization, from photography to film to television, may be quickly implicated in the panoptical surveillance of repressive social order and in the general instrumental rationality of late modern times.[4] It need not be denied that modern optical technology may be used in unacceptable ways, however, to suggest that closer consideration of the second alternative, surface, the fundamental physicality of the optical and visual, might yield a powerful antidote to these abuses.

The importance of architecture to the beginnings of Italian Renaissance perspective—Brunelleschi's plans and elevations, Alberti's "window" overlooking a "pavement"—might have contributed to the crucial, all-important addition of measure and ratio to ancient *skenographia*. And perhaps the extension of optics—or of "visual lines"—to the scale of cartography and astronomy was transitional to the formulation of what I have called *metaoptical* space. The distance traveled by light from far-away mountains or from the stars can be measured in relation to a universal notional two- or three-dimensional framework, and within and against this pure extension all phenomena may be plotted and measured. Any point or movement might be plotted in relation to any other, any point on the earth's surface to any other, any star to any star.

Whether or not we are interested in the history of perspective as a representational construction, the history of the *metaphor* of perspective cannot be easily put aside. "Perspective" has become a pervasive and frequent metaphor, used as if its meaning was shared in distinctly modern discourses of many kinds, from interpersonal and international relations to historical and literary interpretation. Perspective is in fact more than a metaphor: it is now a common epistemological and practical model, with very definite features—a subject at a point, an object, and a definite distance between them. To be sure, the metaphor of perspective has culturally specific roots, but it is also fair to say that as it has gained currency, it has also overflowed its cultural banks.

From the beginning, Italian Renaissance perspective had two contradictory associations—with truth as well as with illusion. Perspective was "true" in the sense of making possible the "right" representation of forms and most importantly the representation of their spatial relations with respect to a viewer. "You know," Alberti wrote, "that nothing painted may ever seem like the true when there is not a certain distance for seeing it."[5] For some 250 years after Brunelleschi's death, however, painter's perspective exemplified illusion and such trickery as anamorphosis. In sixteenth-century Florence, for example, a certain Guglielmo, called Il Giuggiola (that is, sweet fruit of the jujube tree), in what might be called an outburst of popular Platonism, wrote a carnival song about *simulatori* (hypocrites and deceivers), concluding that the whole world is done "*in prospettiva*"—that is, the whole world is a deception.[6] It was not until the classical period of early modern natural science, the early eighteenth century, that the full metaoptical implications of universal opticality were drawn in Newton's "absolute" space and time, which in his *Opticks* he called the "sensorium of God."[7] The infinite extension of coordinate space was in effect the theater of God's thought, proceeding according to a single universal and absolute time.

The German philosopher and mathematician G. W. Leibniz, coinventor with Newton of the calculus, rejected Newton's ideas and in doing so turned the idea of perspective to metaphysical purposes of quite a new kind. His arguments are difficult and fall quaintly on the modern ear, but at the same time they are strangely prophetic of what was to come.

In order to make sense of Leibniz's "perspective," we must briefly consider the history of the idea of the "subject." Martin Heidegger (whom I will discuss in more detail shortly) recognized that the "subject" is a Western idea and that its primary ancestry is Aristotelian. According to Heidegger, a crucial transformation took place in late antiquity when Greek philosophical language, still aglow with something of its ontological authenticity, was translated into Latin. He must have had in mind an author like Martianus Capella, who, paraphrasing Aristotle's *Categories* (and thus the beginnings of logic), wrote toward the middle of the fifth century C.E. that "Everything that we say is either a *subject* in itself, or *of a subject*, or *in a subject*, or both *of* and *in a subject*. A subject is itself not attributed to anything else in such a way that it is dependent on it for its existence, but other things are attributed to it."[8] The translation of Aristotle's *hypokeimenon* as *subiectum* is ostensibly literal, but for Heidegger it marked the beginning of "the rootlessness of Western thought."[9] The ontological core had been lost, with disastrous consequences mounting to the present.

Still according to Heidegger, the meaning of the human "subject" was fundamentally realigned by Descartes as part of the mathematical and mechanical reformulation of the world. In our present terms, the problem was formulated (but not solved) by in effect making the subject a geometric point representing one substance (immaterial mind, the uniquely human soul, the famous *cogito*), "in" another substance, extension, the basis of the material world (including its own body). But if the human being is a union of material body and immaterial mind (as Descartes believed), how can soul and body interact? This is Descartes' equally famous dualism, and the subject might now be said to be doubled.

Descartes proposed a quasi-mechanical interaction according to which soul acts and reacts through the pineal gland in the center of the brain, communicating with its body via the traditional semimaterial spirits.[10] In Leibniz's account, souls are "monads," pure, immaterial, and again pointlike principles of activity, part of a nonspatial reality of their own. Their primary activities are appetition and perception; they are by nature "representative," meaning that each responds to and *represents* to itself the active reality of the monads with which it coexists. Each is a "perpetual living mirror of the universe." A single city, Leibniz continues, if seen from different sides, might seem to have been multiplied *perspectivement* and in effect to be many cities; similarly, because of the infinite multitude of monads (*substances simples*), it might seem that there are different universes, when actually, just as there is only one city, there are only perspectives with respect to one universe according to the point of view of each monad.[11]

This example is not as straightforward as it seems, and Leibniz has made point of view subjective in an entirely new sense. Perspective is no longer optical, it is *spiritual*, active and projective, and inseparable from the essential activities of the human subject. It is vision in an entirely new sense: perspective is now fused with appetite, the living temporal principle giving direction to ongoing representation. The conceptual space opened up by the general and apparently neutral question of appetite—itself a broad and suggestive metaphor—initiated the series of ever deeper temporal principles, such familiar forces as "will," the "unconscious," or "desire."[12] In the continual search for primary motivations, "subjectivity" and vision began to assume their modern forms as well as their modern complexity.

Leibniz might have folded connotations of artifice and illusion into his projective spiritual vision, which was able to rise to the awareness of freedom and to clear and distinct knowledge only in human monads and only in the highest of those. However that may be, there is a deep aesthetic element in Leibniz's scheme. All of these points of view give "variety" to the universe as a whole composition, and to explain and justify this variety, Leibniz appealed to another of the "inventions of perspective," anamorphosis, in which "certain beautiful designs seem to be only confusion, until one finds their true point of view, or until one looks at them by means of a certain glass or mirror."[13] In Augustine's theodicy, which this seems to echo, the world is in reality a great poem set out with antitheses, its apparent oppositions resolved only in the eyes of God.[14] For Leibniz, the "deformities" of our little worlds are united in the beauties of the great world, and God, standing at the true point, with the right glass or mirror, sees an infinitely perfect and universal unity. This anamorphosis made right is the alternative to Newton's "sensorium of God," and it is also the vision that opened Leibniz's "best of all possible worlds" to Voltaire's scorn.

Immanuel Kant offered a reconciliation of Leibniz and Newton, defining space as the "pure form of outer intuition."[15] Significantly, however, he did not describe it in terms of perspective (much less of appetite). Whatever the individual perception of space might be, it was finally, when understood clearly, Newtonian, installed together with Newtonian time (the "pure form of inner intuition") in the transcendental architecture of the pure reason of the human subject.

As part of the lament over the Western tradition already encountered in his history of the subject, Heidegger argued that Kant had "recoiled from" the conclusion that the transcendental imagination—the formative spatiotemporal capacity of the mind in the immediate, sensationary intuition of the world—is radically temporal,[16] leaving Heidegger with his own never-completed philosophical project, *Being and Time*. In fact, Heidegger took his position in a highly developed tradition. It is enough to recall the fundamental importance of temporality in writers as different from one another as Hegel, Schopenhauer, Marx, and Darwin, and by the end of the nineteenth century, space, as a static principle of universal order and rationalization, could be seen to be antithetical to time. Perspective as the geometric construction of space was also the geometric construction of vision, which was demoted and "denigrated" precisely because it had come to be integral with rationalization.[17] In the long earlier tradition, it had been precisely vision's open border with reason that had been the basis of its dignity among the senses. In the meantime, other more temporal "perspectives," not implicitly correctable by absolute space and more closely linked to appetite, began to emerge.

Henri Bergson opposed temporal "intuition" and "intelligence," investigative and inventive human capabilities, to "intellect," abstract rational principle, which he characterized as spatial. He linked intuition with evolution in his *Creative Evolution*, published in 1907. For Bergson, space was metric and mechanical, and the time of clocks and time zones—the institutionalization of Newtonian time and the progressive coordination of the world—was inauthentic "spatialized" time. Even film, new in the years Bergson wrote, projected an inauthentic time of frozen moments, concealing and denying the deeper, living energies of both human and cosmic innovation.

Bergson was world famous, and not only were his ideas admired, they were also controversial and influential in ways he could not have foreseen. His writing found avid readers among artists bent on the creation or justification of characteristically modern forms,[18] and the prospect of intuition (including artistic intuition) as an authentic agent of "creative evolution" gave entirely new importance and license to the aesthetic at the same time that it obviously undercut academic discipline (again perhaps best exemplified by perspective). Whether or not Bergson is any longer much read, the opposition of intuition and reason has shaped modern Western ideas about art and about the value of the institution of art to the present.

However true it may be that the hard-edged, quantitative, and mechanical rigor of modern production, commerce, communication, and war has yielded countless sorry results, the alternative and corrective offered by Bergsonism to this "spatial" culture—a deeper intuitive, "living" temporality—was to have its own dark and murderous blooms in the twentieth century as "creative evolution" merged with existing historicisms (versions of Hegelianism and theories of inevitable progress) or with "primitivism" and the enactment of national and racial destinies. The critique of technological modernity in Heidegger's writings is in basic respects closely parallel to Bergson's and has provided a more authoritative base for a broad array of cultural criticism.[19] It is by no means clear, however, that the ontological reawakening Heidegger tirelessly sought was as universal as the word "ontology" might imply or that it avoided the ethical and political pitfalls to which Bergsonian intuitionism was also liable.

In summary, as perspective became a modern Western metaphor for human being-in-the-world, it underwent a crucial change when Leibniz spiritualized it in linking it to the Cartesian absolute subject; Kant made this absolute subject *transcendental* and its spatial intuition Newtonian. After Kant, the understood nature of subjectivity steadily changed, the traditional view of human rationality and freedom yielding to one or another essentially temporal and prerational intuition or unconscious drive. While perspective as a practical means of representation suffered from its association with metaoptical order and its vast institutionalization, something like "perspective" as the essential representational activity of the subject persisted through these changes in the understanding of the subject itself. Following a familiar pattern, spiritual "perspectives" expanded from the individual and monadic to collective, national, and racial "worldviews" and zeitgeists to take their places in the discourses of cultural style that have figured so prominently in the history of art.

What if all of these "perspectives" were acknowledged to be metaphors, based at one or another level of abstraction and amplification on the fact that the physical process of vision is partially analyzable in geometric terms? What if—to return to the arguments of this book—a "point" of view was acknowledged always to be a surface, in principle a *particular* surface (or two particular surfaces), inseparable from the intricate organism of a living body, the unique and determinate means by which we *react* to the world but also the means by which we *enact* whatever might direct or drive us? For the present, it is enough to say that very useful distinctions and clarifications might be made: optics might be separated from artificial perspective, which might in turn be found to stand in various relations to vision. Vision, to which much of our cerebral capacity is devoted, might be found to have a great many innocent and positive uses, and cultural constructions of vision could be investigated without risk of reduction to one or another quasi-spiritual "perspective" or symbolic form.

Since he thought that only souls are truly alive, Descartes could compare a corpse to a broken clock, and he was very impressed by hydraulic automata he saw in the grottoes of royal gardens. His examples may seem archaic and clumsy, but the metaphor of the body/machine as a hypothesis has been endlessly refinable and correctable at the same time that understanding of the nature of the "machine" and of its physical constituents and constitution has become more and more complex. Leibniz also compared the body to an automaton, but, interested as he was in microscopy (a newly developed part of optics), he saw organisms not as simple machines like clocks but as a system of living machines within living machines to a theretofore invisible depth and complexity of which only God could know, insisting throughout that the organizing and active principle could not itself be mechanical.[20] Leibniz's aggregate monad is of course also an early essay in the characterization of the human body, the human mind, and their interrelations. But it is a step in a historical process by which the terms of Descartes' dualism may be seen to have drawn closer and closer together. As the understood physical complexity of human body and mind (and the understood complexity of the physical itself) has increased and deepened, the absolute particularity of any individual human being has also come more and more clearly into view. The open-endedness of this ongoing historical process does not mean that inferences cannot be drawn from it, and it has already firmly established the basis for the *material subject*, a conception at once of the commonality and of the absolute particularity of every human being. This particularity should provide a basis of value able to stand irrespective of culturally specific understandings of the individual's ultimate origin, nature, or destiny. The *material subject* is a culmination of the physical and physiological project in which optics played an indispensable early part, and it is in itself one of the great achievements of the modern West. It should be acknowledged as such and should be allowed to assume its full stature and historical significance.

In long traditions, souls, spirits (or any number of other equivalent terms) are individual; more specifically, they are the life principles of individuals. The souls of the dead, we say, have "departed." When Descartes defined the modern subject, it became a point, partly to make it locatable in the world of extension but also to separate it from that world, to make it immaterial and in principle immortal, in line with Christian orthodoxy. The analogy of soul and point is hardly an exact one, however, because, while both may be immaterial, all points are alike but all souls are not. If, on the other hand, the "point" of view is assumed not to be geometric but rather a living surface in physical relation, this at once entails particularity, and a sufficient account of an infinitely more complex and finally unique particularity may be given from this beginning.

The modern triumph of intromission was much more than a milestone in the narrow history of optics; it placed the subject once and for all in a world of active physical forces and unified the relation of the subject to the broader world at the basic level of sensation. Vision now meant that the subject was contacted by—was literally "in touch with"—everything visible, and this contact through the activity of light, although contact from a distance, meant that an essential difference between sight and the other senses had been removed.

Intromission, and the more general definition of sensation as response to forces, implies passivity on the part of the subject, who may now be said to be simply "subject" to them. This "passivity," however, is richly layered and structured, made up of positive capacities to react, evolutionarily given adaptations. These adaptations represent an equilibrium achieved between force and counterforce, between environment, ongoing events, and basic capacities (including such emergent modern entities as the unconscious), made evident in reactions but also in characteristically human actions. To keep to our example, the eyes are fitted not only to a certain range of the energy of light but also to the whole attentive life and activity of an individual human being. Seen in these terms, individual "point of view" is the locus and origin of *anthropomorphic* experience and thus of the infinitely more complex spatiotemporality of individual human lives.

By calling experience "anthropomorphic" (referring to humankind taken altogether), I mean that our experience is inevitably and cumulatively human shaped. It is shaped first by the conditions of our given physicality, then by the various social spaces we inhabit—that is, by the cultures in which we take our places and continue to enact, maintain, and transform. Perhaps the material subject provided the metaphorical core for the relation of the individual to the metaphorical "forces" of modern life—economic, demographic, and historical—in which case it further implies or demands a just equilibrium between individuals and these forces.

The founders of modern Western science rejected anthropomorphism, the assumption that the natural world is like ourselves, and abandonment of this rejection has opened the natural world to undreamed-of understandings and practices. This systematic estrangement of ourselves from nature—of which we are still nonetheless inevitably part—has in fact finally relocated us in a new nature within which there is unprecedented human control together with vast multiplications of human powers and alternatives. In effecting these powers and choosing among these new options, it is imperative to recall that if nature is not anthropomorphic, we ourselves are and that this always individual human form is in itself a touchstone of value.

Leibniz already recognized that a single city (or universe) might seem to be many from many points of view, and as he defined his "perspectives," no one could simply occupy the "point of view" of another. This brings us to perspectivism, a peculiarly modern form of skepticism. It is not necessary to defend the idea that points of view are interchangeable and the visible world "objective" in the sense that everyone may see just the same thing in order to reject the radical perspectivist position that each point of view constitutes the world taken altogether, a position most often associated with the late Romantic hyperaristocratic megalomania of Friedrich Nietzsche. Given the material subject, however, all points of view are necessarily grounded in human corporeality, then in the *specific history* of a corporeality. Skeptical perspectivism must deny this commonality and so deny the common basis for the specific histories of human lives.

Perspective is often associated with "objectivity,"[21] but then "objects" exist for subjects (as Heidegger also pointed out), and common usage—still distinctively modern Western usage—has incorporated perspective in more pragmatic forms to refer to the ways in which objects, problems, things "thrown in the way" may be dealt with. "Perspective" is generally a positive term, and in common usage "points of view" are generally conceded validity. This important sense of "perspective" has entered the realm of common sense and, as I have remarked, has surpassed the limits of its cultural origins. We are, for example, accustomed to "getting other points of view" or "approaches" (an obviously related metaphor). "Perspective" connotes positive coolness and distance, and we "lose perspective" when we are overcome by feeling or passion. "Perspective" may even connote wisdom: we "gain perspective" with experience, and some event might "give us a whole new perspective."

Rather than being absolute, as geometry might imply, perspectives are acknowledged in common usage to be essentially physical and therefore temporal, structured but also provisional (as the language permits us to say). They are actual but also revisable (as the language also permits us to say). By extension, perspectives are also temporal in the sense of being historical. If, for example, we speak of the Danish or Indonesian "perspective" or "point of view," we mean to refer to judgments shaped not so much by personal experience as by the experience of members of specific cultural, social, and political traditions. Such "points of view" are, however, statistical rather than essential. If some Danes or Indonesians disagree with others, this does not undermine point of view but more properly refines it.

The optical—that is, the description of sight as a physical relation, which has led us to the definition of the material subject—is deeper than the metaoptical space abstracted from it. Metaoptical space, far from being merely abstract, is a space of prediction and control, and its degree of abstraction is like that of the energies, the concentrated natural resources that can be made to work through it for human ends. The vast multiplication of human power founded in the metaoptical has created an unprecedented demand for choice, a demand that these human purposes in fact be *humane* purposes. The material, anthropomorphic human subject, which eludes description in metaoptical terms and of which the activities of vision are a major part, provides a standard for judgment relative to which these choices may be well made. The issues that inevitably arise with greater scientific knowledge and technological innovation have much more to do with the *desirability* of new possibilities.

Barbara Stafford placed modern medical imaging in the history of human art, arguing that it is at long last post-Platonic.[22] It does not proceed from the distrust of vision, nor does it seek an ideal. The body is not a corpse carried around by the soul, as the Platonists liked to say. On the contrary, the concern of these images is absolute particularity—previously unrepresentable particularity. Although these images may be moving and even awful, they are not aesthetic, and they are the results of vast, patient, collective, but directed effort.

At the same time, they are the direct descendants of Renaissance drawing and optics, not in subject matter but in the relation between means of representation and subject matter. Visibility, like all sensation, is physical. It is represented physically, and these representations may be participated in physically. It is in such terms that the technological construction and reconstruction of vision have made possible the art of Bill Viola, in which elemental human mortality and spirituality are so intimately united to contemporary visuality.

NOTES

Introduction

1. Vitruvius, *De architectura*, I.1.4. All references to ancient authors in the text and notes are to Loeb Classical Library editions unless otherwise indicated.
2. See Crary, *Techniques of the Observer*; Crary, *Suspension of Perception*.
3. On virtuality, narrative, and fiction, see Summers, *Real Spaces*, 341–42.
4. See Summers, "Heritage of Agatharcus."
5. Aristotle, *Rhetoric*, 1405b.
6. Horace, *Ars poetica*, 361; Lee, "*Ut pictura poesis*."
7. Plutarch, *De gloria Atheniensium*, 3 (Moralia 346F–347A).
8. Summers, *Judgment of Sense*.
9. The opening lines of the pseudo-Aristotelian "On Gorgias," 979a (Aristotle, *Minor Works*, 496–97).
10. Ackerman, "Alberti's Light," 3–4.
11. On demonstration, see Kretzmann et al., *Cambridge History of Later Medieval Philosophy*, 496–517. For Filarete, see chap. 2, n. 87.
12. Aristotle, *Physics*, 194a.
13. On the "middle sciences," see Kretzmann et al., *Cambridge History of Later Medieval Philosophy*, 525; Summers, *Judgment of Sense*, 214 n. 48. See also Farago, "Leonardo's Color and Chiaroscuro Reconsidered."
14. Alberti, *On Painting*, 1. Throughout I will refer to Alberti, *Opere volgari*, where both the Italian and Latin versions are given. See also Alberti, *On Painting and On Sculpture*. These versions of Alberti's texts have a common numeration by sections, and I will simply refer to *On Painting* with the appropriate section number.
15. Cicero, *De amicitia*, V.19. In investigating friendship, the interlocutor says, the ways of the philosophers and sophists should be avoided in favor of the "plumper Minerva" of experience and Nature. "*Pinguis*" has more connotations of health and fecundity than of crudeness and dullness favored by translators of both Cicero and Alberti, and Alberti's use of Cicero's proverbial phrase is in general compatible with the stated aim of his treatise. Although set out in other terms, Kelly, *Leon Battista Alberti*, stresses the interrelations of optics, cartography, surveying, astronomy, and architecture. Grafton, *Leon Battista Alberti*, titles his third chapter "From New Technologies to Fine Arts: Alberti among the Engineers" (71–109). What we call "engineering" was (among the middle sciences) the *scientia ingeniorum* (Summers, *Judgment of Sense*, 254 n. 53).

16. Alberti, *On Painting*, 2. See also Greenstein, "On Alberti's 'Sign'"; Fehrenbach, "Der oszillierende Blick." Alberti's words must recall the line-splitting contest of Apelles and Protogenes; see Pliny the Elder, *Natural History*, 35.xxxvi.81–83; Waal, "*Linea Summae Tenuitatis* of Apelles."
17. Summers, "Pandora's Crown."
18. Landino, *Comento*, 1:234–5.
19. Federici-Vescovini, "Inserimento."
20. For a review of the few works attributed to Alberti, see Alberti, *On Painting and On Sculpture*, 142–54. On the "Barberini Panels," about whose attribution to Alberti Grayson expressed misgivings, see Christiansen, *From Filippo Lippi to Piero della Francesca*, where they are given to Fra Carnevale.
21. Vasari, *Vite*, 3:282–90. Vasari said that "in the house of Palla Rucellai there is a portrait of himself made *alla sfera*, and a panel with very large figures in *chiaro e scuro*." Vasari also described a perspective of Venice and San Marco with figures painted by other masters.
22. Kaufmann, "Perspective of Shadows." The investigation of shadows was mostly astronomical, not optical, and that is why Alberti lists as sources of light "stars, moon, sun and Venus" as well as fires (*On Painting*, 11).
23. Edgerton, *Renaissance Rediscovery*, 92; Peckham, *John Pecham and the Science of Optics*.
24. Kelly, *Leon Battista Alberti*, 76. On the horizon, see chap. 1, n. 58.
25. Farago, *Leonardo da Vinci's "Paragone,"* 239.
26. Ibid., 184–85: "This science [of painting] is the mother of perspective, that is, visual lines. . . . [T]he science of visual lines gave birth to the science of astronomy, which is a simple perspective because all of [its elements] are visual lines and intersected pyramids."
27. Ames-Lewis, *Drawing*.
28. Lindberg, "Lenses and Eyeglasses." Gage, *Color and Meaning*, 90–93, illustrates the first representations of reading glasses by Tomaso da Modena from ca. 1352. Lenses are described as an aid to reading by Roger Bacon around 1260; eyeglasses were used in Tuscany and northern Italy toward 1300 but may have been used earlier in the thirteenth century.
29. See Burton, "Optics"; Smith, *Ptolemy's Theory*.
30. Although the works of Plato were not generally current in the Western Middle Ages, Chalcidius's fourth-century translation of the first half of the *Timaeus* made available Plato's account of vision. See Lindberg, *Roger Bacon and the Origins*, xxix–xxx. See also Macrobius, *Saturnalia*. Among those in Byzantium with a specialized interest in optics, extramission seems also to have prevailed. See Nelson, "To Say and to See." As might be expected, Marsilio Ficino held out for the old Platonic theory, adapting

it to his explanations of both love and magic; see, for example, Ficino, *Commentary*, 159–61. On Ficino's interest in optics, see Allen, *Nuptial Arithmetic*, 97–100.

31. Baxandall, *Words for Pictures*, 152.
32. Aristotle might be taken to accept extramission at *Meteorology*, 372a, but he expressly rejected it at *De anima*, 419a18–19, and *De sensu et sensato*, 438a–b.
33. Aristotle, *De sensu et sensato*, 438a.
34. Lindberg, *Theories of Vision*, chap. 4.
35. Kemp, *Science of Art*, 264–84.
36. Aristotle, *De sensu et sensato*, 439a29–32: "color [*chroma*] either is in the limit [of a body] or else is the limit itself. This is why the Pythagoreans called the surface of a body its color." Color is thus closely related to Aristotle's vexed notion of place (*topos*); see *Physics*, IV.I–II.
37. Summers, *Judgment of Sense*, 55; Aristotle, *De anima*, 425a27, 425b7.
38. Aristotle, *De sensu et sensato*, 439b–440a.
39. Aristotle, *Meteorology*, 374b10–15.
40. Gombrich, "Heritage of Apelles." See also Smith, *Ptolemy's Theory*, 122.
41. Alberti, *On Painting*, 18. Parronchi, *Studi*, 259, persuasively argued that the "philosopher" in question was Biagio Pelacani da Parma, whom I shall discuss at some length in chapter 2.
42. On Alberti and Nicholas Cusanus, see Harries, *Infinity and Perspective*.
43. Alberti, *On Painting*, 52.
44. Baxandall, *Giotto and the Orators*; Baxandall, *Painting and Experience*.
45. Alberti, *On Painting*, 52.
46. Pliny the Elder, *Natural History*, 35.xxxvi.60. Ghiberti, *Lorenzo Ghiberti*, 68, wrote that Apollodorus "began to brighten (*chiarire*) luminous bodies and to show by what natural reasons light might be given to painted things"; his were the first paintings "to display any virtù."
47. This issue was recently brought to the fore by Hockney, *Secret Knowledge*. Hockney's arguments are in important respects historical, and some may stand up to scrutiny better than others, but this does not affect the validity of his general position.

Chapter One

1. This is the opening thesis of Gombrich's *Art and Illusion*. This book, based on Mellon Lectures given at the National Gallery in Washington in 1956, was first published in 1960.
2. The Paleolithic paintings at Chauvet display modeling as well as another optical feature, overlapping, but in these respects they are striking in their differences from later painting.
3. Aristotle, *De anima*, 418a27–419a25.

4. Gombrich, "Heritage of Apelles."

5. Moreno, *Apelles*, 99–100.

6. See Pollitt, *Art of Ancient Greece*, 147, 271; Pollitt, *Ancient View of Greek Art*, 217–24. On Agatharcus of Samos, see Hobey-Hamsher, "Agatharcos"; on Apollodorus of Athens, see Hobey-Hamsher, "Apollodoros." See also Lydakis, *Ancient Greek Painting*, 128–30. Lydakis sees the influence of Apollodorus in paintings like the so-called *Knucklebone Players* and white-ground *lekythoi* with modeled forms. Modeling is very light, confined to drapery in representations of women.

7. See, for example, Behl, *Ajanta Caves*, 78–79.

8. For an introduction to Thales, see Hornblower and Spawforth, *Oxford Classical Dictionary*, 1491.

9. See Erlande-Brandenburg et al., *Carnet de Villard de Honnecourt*, pl. 40.

10. Pliny the Elder, *Natural History*, 35.v.15. Aristotle, *Topica*, 140a20–23, wrote that definitions must be clear enough to imply their contraries. If they are not clear, they will be like the works of early painters: "unless they were inscribed, it was impossible to recognize what each figure represented."

11. Pliny the Elder, *Natural History*, 35.xliii.151; Stoichita, *Short History of the Shadow*, 11–41; Muecke, "'Taught by Love.'"

12. Farago, *Leonardo da Vinci's "Paragone."*

13. Aristotle, *Poetics*, 1449b5.

14. Ibid., 1450a9–12.

15. Ibid., 1450b28.

16. Vitruvius, *De architectura*, VII, preface, 11.

17. Aristotle, *Poetics*, 1449a, 18–19. Brown, "Three and Scene-Painting Sophocles," argued that the crucial verbless phrase—the title of his paper—was not by Aristotle but was a later addition and examined a number of tragedies for references to painting considerably more illusionistic than it is reasonable to expect, concluding that the existence of such painting is unlikely. If Aristotle's text can be dismissed, then Vitruvius may become the better source, perhaps suggesting an earlier date. Pliny the Elder, *Natural History*, 35.xl.143, in a list of minor painters (including a painter of *scaenae*), cited a certain Simonides, who painted *Agatharcum et Mnemosynen*. As pointed out by Robertson, *Greek Painting*, 176, this might refer to two paintings or one, and in the second case Agatharcus would have been shown with the mother of the Muses. Crowning by a divine personification was a common form of tribute in Greek art, and "one would like to think that Simonides was one of the artists who developed the theory of perspective, and painted such a picture in piety to the first master."

18. Plutarch, *Pericles*, xiii.2.

19. See Plato, *Critias*, 107c–d.

20. See Sinn and Wehgartner, *Begegnungen mit der Antike*, cat. no. 51, p. 116, abb. 9, p. 179.
21. The Renaissance humanist Pomponius Gauricus, *De sculptura*, 127, 261, noted that Demosthenes refers to Agatharcus as *grapheus*, which usually means "writer." Gauricus's point is that painting and writing are the same, but Demosthenes might have meant that Agatharcus was in fact both painter and writer, providing some slim further evidence for his commentary. Alcibiades locked Agatharcus in his house to force him to make paintings there. In some versions, Agatharcus escaped; in others, he grievously insulted Alcibiades by seducing his concubine. See Demosthenes, *Against Meidias*, 174–75, 362–63.
22. Vitruvius, *De architectura*, VII, prooemium. This text is also controversial and has been variously translated. The crucial part of the text (which also implies the scheme taken up by Democritus and Anaxagoras) reads "*ad aciem oculorum radiorumque extentionem certo loco centro constituto, ad lineas ratione naturali respondere.*" See Panofsky, *Perspective as Symbolic Form*, 100–102; Pollitt, *Ancient View of Greek Art*, 234–36, which for the most part follows J. White, *Birth and Rebirth of Pictorial Space*, 248–62. See also Richter, *Perspective in Greek and Roman Art*, 58–61. Richter concluded that if philosophers may have formulated a consistent one-point construction, painters did not use it, and that consistent constructions such as those in figure 1.4 "may be due to the correct observation of an individual artist" (52–53).
23. See Summers, *Real Spaces*, chap. 6, on *skiagraphia* and optical planes.
24. For a review of the question and an argument that the term refers to the simultaneous contrast of colors, see Keuls, "*Skiagraphia* Once Again." This argument was vigorously rejected by Pemberton, "Note on *Skiagraphia*," who associated *skiagraphia* primarily with value contrast, in imitation of Phidian sculpture. I have followed this second argument. See also Keuls, *Plato and Greek Painting*.
25. Quintilian, *Institutes*, VIII.v.26.
26. Stoichita, *Short History of the Shadow*.
27. Homer, *Odyssey*, X.495, XI.207.
28. Plato, *Republic*, 509e–510.
29. Pliny the Elder, *Natural History*, 35.xxxvi.
30. Ibid., 35.xi.29.
31. Bruno, *Form and Color*, 23–30.
32. Plutarch, *De gloria Atheniensium*, 2 (*Moralia* 346A).
33. Quintilian, *Institutes*, XII.x.4. Alberti (*On Painting*, 44) would follow Quintilian.
34. Pliny the Elder, *Natural History*, 35.xxxvi.61.
35. A similar scheme is to be seen in the *Lion Hunt* mosaic from Pella (Lydakis, *Ancient Greek Painting*, 214–15).

36. Pliny the Elder, *Natural History*, 35.xxxvi. On "scattered beauty," see Summers, *Michelangelo*, 186–99.

37. Quintilian, *Institutes*, XII.x.5.

38. See Keuls, *Plato and Greek Painting*, 9–32. Keuls argued that Plato was not hostile to painting or even very interested in it but also said that "the sense of sight and the one art which relates exclusively to vision, painting, furnish Plato's favorite metaphors for everything which he considers relative, transitory and deceptive" (47).

39. Plato, *Theaetetus*, 208e.

40. Plato, *Republic*, 596d–e.

41. Ibid., 514–18.

42. Plato, *Sophist*, 234b.

43. Plato, *Protagoras*, 356d.

44. Plato, *Republic*, 585a.

45. Ibid., 586b–c, 602c–d.

46. Dionysius of Halicarnassus, *Isaeus*, 4, in Russell and Winterbottom, *Ancient Literary Criticism*, 178–81.

47. See Horace, *Ars poetica*, 1–13; Vitruvius, *De architectura*, VII.v.3–7.

48. Gombrich, *Preference for the Primitive*, 19–22.

49. Plato, *Republic*, 530d. Human uprightness is also "framed" to the contemplation of the heavens. See Summers, *Michelangelo*, 576 n. 22.

50. Aristotle, *Metaphysics*, 1024b23.

51. See n. 22.

52. Gombrich, "Light, Form, and Texture," 19.

53. Pliny the Elder, *Natural History*, 35.xl.126–27, 35.xxxvi.76–77.

54. Keuls, *Plato and Greek Painting*, 139–50, reviewed the question of Pamphilus's system of education and of Plato's possible objections to it. She noted Plato's hostility to Democritus (one of the commentators on Agatharcus's treatise on *skenographia*) and suggested a connection between Democritus and Pamphilus.

55. Moreno, *Apelles*; for Helen of Egypt, see Pliny the Elder, *Natural History*, 35.xxxii.50.

56. Pliny the Elder, *Natural History*, 35.xxxiii.50; Bruno, *Form and Color*, 53–59.

57. Moreno, *Apelles*, 16–18. See also Voisin, "Exoriente sole." I am grateful to Malcolm Bell III for bringing this article to my attention.

58. Macrobius, *Saturnalia*, 1:448–52.

59. Pliny the Elder, *Natural History*, 35.xxxvi.92, 95–96.

60. Ibid.

Chapter Two

1. Lindberg, *Roger Bacon and the Origins*, xxvii–xxviii.

2. Lindberg, *Theories of Vision*, 49.

3. For the Latin, see Alhazen, *Opticae thesaurus*; for the Arabic with an English translation, see Alhazen, *Optics of Ibn al-Haytham*; for an Italian translation of Alhazen known in the fifteenth century, see Federici-Vescovini, "Problema."
4. Parronchi, *Studi*, 3–90.
5. Panofsky, *Perspective as Symbolic Form*.
6. For an introduction to the question of the arbitrariness of perspective as a system of representation, see Gombrich, *Art and Illusion*, 242–87 (see also xxv–xxxvi, Gombrich's new preface to this edition); Goodman, *Languages of Art*, 10–19; Gombrich, "The 'What' and the 'How.'"
7. Trachtenberg, *Dominion of the Eye*.
8. On metaopticality and modernism, see Summers, *Real Spaces*, chap. 7.
9. Hills, *Light of Early Italian Painting*, 64–71.
10. Alhazen, *Opticae thesaurus*, II.13.
11. Aristotle, *De anima*, 431a14–18.
12. On Brunelleschi's machines, see Praeger and Scaglia, *Brunelleschi*.
13. Manetti, *Vita di Filippo Brunelleschi*, 4–44.
14. Borsi and Borsi, *Paolo Uccello*, 349–51.
15. Goldthwaite, *Building of Renaissance Florence*, 54–56.
16. For parallel Italian and English texts, see Manetti, *Life of Brunelleschi*, 52–55.
17. Vasari, *Vite*, 3:143–44.
18. For Leonardo's phrase, see Fumagalli, *Leonardo*, 38.
19. Alberti, *On Painting*, preface.
20. Hyman, *Brunelleschi in Perspective*, 24, figs. 1 and 2.
21. Tanturli, "Rapporto del Brunelleschi."
22. Manetti, *Life of Brunelleschi*, 42–43: "also in those times [of his youth], he himself demonstrated that which painters today call *prospettiva*."
23. Edgerton, *Renaissance Rediscovery*, stressed the importance of medieval optics and of the availability of Ptolemy's *Geographia*, both of which, however, might point to an earlier date. Parronchi, *Studi*, 242–43, placed Brunelleschi's demonstration around 1424, connecting it closely with the return to Florence of Paolo Toscanelli in the previous year. According to Manetti, Brunelleschi and Toscanelli were friends for "more than 40 years" (*Life of Brunelleschi*, 55–57), which would put the beginning of the friendship close to 1400.
24. See Goffen, *Masaccio's "Trinity."*
25. Manetti, *Life of Brunelleschi*, 42–43.
26. Summers, *Real Spaces*, 511–17. See n. 57. Lang, "Brunelleschi's Panels," argued for the importance of Brunelleschi's demonstrations as architectural drawing but stressed this drawing's actual theatrical uses, which of course might have been interesting to Brunelleschi as well. A small silver relief in the Louvre has been attributed

to Brunelleschi (Parronchi, *Studi*, fig. 92b) but also associated with Alberti (Ceriani in Christiansen, *From Filippo Lippi to Piero della Francesca*, 119) and with a number of other major Florentine artists of the first half of the fifteenth century (Kemp in Levenson, *Circa 1492*, 240, leans toward Filarete). In addition to its importance in the history of painter's perspective and central plan architecture, the building in the background might be regarded as a reconstruction of a Vitruvian *scaena* (Vitruvius, *De architectura*, V.vi.8, V.vi.3). On the size and shape of Brunelleschi's panel and mirror, see the careful analysis of Degl'Innocenti, "Dimensionamento della tavoletta." Degli'Innocenti considers a horizontal rectangular panel (and mirror) a better match for Manetti's description, but in other respects, his argument is close to that given here.

27. Vitruvius, *De architectura*, VII, preface, 11.
28. This principle is often repeated. See Alhazen, *Opticae thesaurus*, I.24 (p. 7); Alhazen, *Optics of Ibn al-Haytham*, I.3.1 (p. 13).
29. Ghiberti, *Lorenzo Ghiberti*, 64.
30. Ibid., 305. For the text paraphrased and elaborated by Ghiberti, see chap. 1, n. 22.
31. Federici-Vescovini, "Prospettiva del Brunelleschi." See also Parronchi, *Studi*, 238–39. Klein, "Pomponius Gauricus on Perspective," 102–4, stressed Pelacani's importance for the adaptation of optics to painting in both northern and central Italy.
32. For a survey of literature and introduction to Biagio Pelacani da Parma, see Raynaud, "Perspectiva naturalis."
33. Vasari, *Vite*, 3:159–60.
34. Praeger and Scaglia, *Brunelleschi*, 118, 143–44. For an introduction to Giovanni di Gherardo da Prato, called L'Acquettino, and his objections to Brunelleschi's dome, see Camerota, *Nel Segno di Masaccio*, 37. Gherardi is also implicated in Manetti's *La Novella del Grasso*, set in Florence in 1409–10; see Manetti, *Vita di Filippo Brunelleschi*, 13 n. 1.
35. Federici-Vescovini, "Questioni di 'Perspectiva,'" 163–243.
36. Gherardi da Prato, *Paradiso degli Alberti*, 67: "leggiadre della dilettevole prospettiva dimostra."
37. Ackerman, "Alberti's Light," 4.
38. Gombrich, "From the Revival of Letters to the Reform of the Arts." In 1413, Niccoli was ridiculed for measuring and reconstructing ancient buildings (424). Such activity would have yielded inscriptions and since Niccoli is also said to have measured the height of obelisks must have involved surveying of the kind discussed in chap. 1, nn. 8–9.
39. Alberti, *On Painting*, 5; Vitruvius, *De architectura*, VI.2.3, where it is argued that, whether simulacra come into the eye or rays are emitted by it, we may be deceived by sight in the same way.

40. Alberti, *On Painting*, 6.

41. Lindberg, *Theories of Vision*, 178–208. Lindberg stressed that Kepler's account was the culmination of the project begun by Alhazen.

42. Galen, *Galen on the Usefulness of the Parts of the Body*, 479. Galen also compared the crystalline humor to a mirror.

43. "In principle" because although the geometric center of the eye governs its structure and action, it is a pattern on a surface that is transmitted from pupil to crystalline humor to optic nerve.

44. Alhazen, *Opticae thesaurus*, I.4.4 (p. 3), I.6.33 (p. 21).

45. Aristotle, *De sensu et sensato*, 438b8–10; see also [Plato?], *Alcibiades I*, 132d–133a.

46. Alhazen, *Opticae thesaurus*, IV.2 (p. 102), referring to I.1, 2, 3, 14, 18, 19 (pp. 1–22).

47. Quoted in Lindberg, *Theories of Vision*, 49.

48. Ibid., 56: "it may even mark the termination of the optical phase of this transmission [of visible forms]."

49. See Anzulewicz, *De forma resultante in speculo*, 180–211, esp. 187, 195; Caparello, *Senso e interiorità*, 73, citing Albertus's commentary on Aristotle's *De anima*, l. 2, tr. 3, cap. 15, 1.4–5: "Scias autem, quod sicut fit species visibilis in oculo, ita fit in speculo, quod est tersum politum recipiens formam et repraesentans iterato, sicut est res." Caparello noted that at *De anima*, l. 3, tr. 1, cap. 2, 1.82–86, Albertus called *imaginatio* a *speculum animatum*: it is higher and more abstract than sense because it can present images of the absent, but it is a passive faculty (like sight) in comparison with the *estimativa*, which judges particulars, thus to become a *cogitativa*, or particular reason. This is a fusion of the greatest importance. References are to Albertus Magnus, *Opera omnia*, VII.I, *De anima*.

50. Aristotle, *De anima*, 431b: "for the stone does not exist in the soul, but only the form of the stone."

51. Dante Alighieri, *Convivio*, III.ix.7–8.

52. For a review of the question, see Grafton, *Leon Battista Alberti*, 354 n. 2; Rykwert and Grayson (in Rykwert and Engel, *Leon Battista Alberti*, 21, 31) concurred, stating that the Latin version was written first, the Tuscan between one and two years later.

53. For Alberti's famous and much-maligned comparison of a painting to an open window, see *On Painting*, 19.

54. Manetti, *Life of Brunelleschi*, 44–45.

55. Ibid. On the fate of the panel, see Manetti, *Vita di Filippo Brunelleschi*, 57, with valuable commentary on the text.

56. Vitruvius, *De architectura*, I.ii.2. The kinds of *dispositio* (in Greek, *ideae*), are plan (*ichnographia*), elevation (*orthographia*), and projection (*skenographia*).

57. Krautheimer and Krautheimer-Hess, *Lorenzo Ghiberti*, 229–53; Vasari, *Vite*, 3:142:

"levarla con la pianta e profillo e per via della intersegazione." Although usefully approximate, Krautheimer's diagram of the construction resulting from Brunelleschi's use of plan and elevation drawings is incorrect. Numbers 2 and 4 should correspond to the breadth of the front face of the Baptistery, and 3 on the plan, the line of sight, was omitted. Numbers 1 and 5 should locate buildings at the sides of the piazza, which have been unaccountably placed on the ground line rather than the optic plane ending at the horizontal *b*. Unfortunately, these errors were not corrected in Summers, *Real Spaces*, fig. 288.

58. Manetti, *Life of Brunelleschi*, 45.

59. Ibid.

60. Farago, *Leonardo da Vinci's "Paragone,"* 182–83.

61. Olson, *Florentine Tondo*, 97.

62. Ibid., 97 n. 79.

63. Schwarz, "Mirror in Art."

64. Alberti, *On Painting*, 28. In the Loeb Classical Library edition of Pliny (35. xl.147–48), the painter to whom Alberti refers is Iaia of Cyzicus, *perpetua virgo*, who painted when Marcus Varro was young. Her portraits, mostly of women, were highly prized, and she painted her own portrait using a mirror. Perhaps Alberti had read a manuscript of Pliny in which her name was Maia, which, recalling Boccaccio, became Marcia. See also Buettner, *Boccaccio's "Des cleres et nobles femmes,"* 46–49.

65. On Alberti's "pavement," see chap. 3, n. 33.

66. Filarete, *Treatise on Architecture*, 304–5 (178r–179r).

67. Ibid.

68. Parronchi, "Le due tavole prospettiche del Brunelleschi," in *Studi*, stressed the importance of the geometry of the plane mirror and of reflected rather than direct vision for Brunelleschi's first demonstration. He also reviewed arguments for Brunelleschi's having simply painted on a mirror, rejecting them in favor of construction, as I have done here. Edgerton, *Renaissance Rediscovery*, 145–52, argued that Brunelleschi set panel and mirror side by side, using the mirror reflection as a model for his painting, citing Dante's comparison of the "pupil" (which Dante apparently understood to mean the crystalline humor) to a lead-backed mirror. See n. 51.

69. Alhazen, *Opticae thesaurus*, I.25 (p. 15).

70. Ibid., I.26.

71. Ibid., II.3 (p. 25). If we follow Grafton, *Leon Battista Alberti*, 145, in supposing that Manetti's biography records Brunelleschi's own responses to much earlier circumstances, then it may be significant that Brunelleschi compared the opening on the front of the Baptistery panel to a lentil in size and shape because the opening itself might be an image of the crystalline humor.

72. Alhazen, *Opticae thesaurus*, II.3 (p. 25).

73. Witelo in Alhazen, *Opticae thesaurus*, III.4 (pp. 85–87).
74. See Takahashi, *Medieval Latin Traditions*, 39–55, for the variant texts and the issues surrounding the argument. Takahashi noted that Aristotle was considered in antiquity to have made great contributions to optics and concluded that the law of reflection seems to have been formulated between Aristotle and Euclid. In his *De sensu et sensato* (438a9–10), Aristotle excused Democritus's theory of vision because there was no theory of mirrors and reflections when he wrote, implying that Aristotle himself had such theory. See also Alhazen, *Opticae thesaurus*, V.11 (p. 132); Witelo in Alhazen, *Opticae thesaurus*, V.49 (p. 212).
75. Alhazen, *Opticae thesaurus*, V.49 (p. 212).
76. See n. 49.
77. Peckham, *John Pecham and the Science of Optics*, 181.
78. Parronchi, *Studi*, 583–641; Toscanelli, *Della prospettiva*.
79. See Garin, *Cultura filosofica del Rinascimento italiano*, 313–34.
80. Manetti, *Life of Brunelleschi*, 55–57.
81. In 1475, Toscanelli was given permission to make an opening in the lantern of the dome of Florence Cathedral in order to establish a partial *meridiana* on the floor below. See Heilbron, *Sun in the Church*, 70–71.
82. Lucretius, *De natura rerum*, 414–19.
83. Toscanelli, *Della prospettiva*, 36:8v–9r.
84. Ibid., 41:13v.
85. Ibid.
86. See n. 34.
87. Filarete, *Treatise on Architecture*, 305 (179r).
88. Peckham, *John Pecham and the Science of Optics*, 123–24; see Kemp, "Leonardo and the Visual Pyramid."

Chapter Three

1. Dhanens, *Hubert and Jan van Eyck*, 40–43.
2. Consistent one-point construction has long been treated as a fundamental sign of difference between the Renaissance in the north and south of Europe. See Panofsky, *Early Netherlandish Painting*, 1:1–20. Such cultural generalizations have prompted attempts to demonstrate an alternative northern system. See Elkins, "On the *Arnolfini Portrait* and the Lucca *Madonna*." Elkins asked, "Did Jan van Eyck have a perspective system?" and answered his own question "No," urging that investigators draw lines more carefully in their analyses. It might also be considered that there are no traces of construction in the underpainting of the *Arnolfini Portrait* and that a plane mirror might have been used as an aid in the making of the painting.
3. Baxandall, *Giotto and the Orators*, 106, 165.

4. Hollanda, *Four Dialogues*, 15–16.

5. See Nuttall, *From Flanders to Florence*, esp. 105–17. See also Borchert, *Age of Van Eyck*.

6. See Hall, *Arnolfini Betrothal*.

7. On Guillaume Dufay, see Planchart, "Guillaume Dufay." Dufay's career paralleled Alberti's fairly closely: Dufay was in Bologna from March 1426 to August 1428, then in the papal chapel from 1428 to 1433, returning in 1435. As part of the pope's entourage, he composed his great motet for the dedication of Florence Cathedral on March 25, 1436. See Wright, "Dufay's *Nuper rosarum flores*." Dufay left the papal chapel once and for all in 1437. His contact in Florence was the cathedral's organist after 1432, Antonio Squarcialupi, considered the greatest player of his time and a close member of the Medici circle until his death in 1480. In his *Vita*, written in 1437 or 1438, Alberti describes himself as one of the foremost performers on the organ and as someone whose criticism was of use to many musicians. See Watkins, "L. B. Alberti in the Mirror," 7.

8. Panofsky, *Early Netherlandish Painting*, 150, 435 n. 197; Alberti, *On Painting*, 26 (see chap. 4, n. 57).

9. Baxandall, *Giotto and the Orators*, 97–111, 163–68; *Dizionario biografico degli italiani*, 44:113–21.

10. Alberti, *Momus*.

11. Baxandall, *Giotto and the Orators*, 165. Writing in 1449, Cyriac of Ancona was more emphatic: Rogier van der Weyden "is held to be the outstanding painter of our time after the distinguished Jan of Bruges" (*Later Travels*, 365).

12. Alberti, *On Painting*, 53.

13. See chap. 1, n. 53.

14. Baxandall, *Giotto and the Orators*, 107, 166.

15. See Fiorani, *Marvel of Maps*.

16. Panofsky, *Early Netherlandish Painting*, 3, 441, 203 n. 6, where it is suggested that the map might have resembled the *Mappamondi Estense*, a Catalan map variously dated after 1450 (see Destombes, *Mappemondes*, 52.12). This map is illustrated by Dhanens, *Hubert and Jan van Eyck*, 171, fig. 120, and although it is round, it is not a grid map. Sterling, "Jan van Eyck avant 1432," 69–77, connected the map with plans for a crusade around 1436 and related it to early-fifteenth-century naturalistic cartography (see Harley and Woodward, *History of Cartography*, 1:486–89). I will discuss Sterling's arguments in more detail.

17. On the basis of this document, the map was dismissed by Paviot, "Mappemonde attribuée a Jan van Eyck."

18. Dhanens (*Hubert and Jan van Eyck*, 170), who did not dismiss the map from van Eyck's work on the basis of the documents cited in n. 17, assigned the map to the

period of Philip's construction of a *studiolo* in Brussels (that is, between 1431 and 1436) and suggested that Fazio might have known of it from someone in the entourage of Alfonso I, who was made a knight in Philip's Order of the Golden Fleece in Ghent in 1445. (The order was founded in 1430.)

19. Ibid., 172.
20. Kent, *Cosimo de' Medici*, 264.
21. Ghiberti, *Lorenzo Ghiberti*, 89.
22. Baxandall, *Giotto and the Orators*, 107, 166.
23. For an introduction to the literature on Philip the Good, see Bourassin, *Philippe le Bon*.
24. Sterling, "Jan van Eyck avant 1432," 28; Dhanens, *Ghent Altarpiece*, 103; Philip, *Ghent Altarpiece*, 180–92.
25. Philip, *Ghent Altarpiece*, 182–83.
26. See Paviot, *Ducs de Bourgogne*. It was Paviot who dismissed van Eyck's putative map from his works. See n. 17; see also Paviot, "Burgundy and the Crusade."
27. Baxandall, *Giotto and the Orators*, 166. Sterling, "Jan van Eyck avant 1432," 72, noted that *regiones* (continents) and *loca* (places) reflect the terminology of Ptolemaic maps, which might mean that Fazio, the mapmaker, or both knew this terminology.
28. Bacon, *"Opus majus,"* IV (1:259–60); Harley and Woodward, *History of Cartography*, 1:322, fig. 18, 28, pl. 36.
29. Leopold, *How to Recover the Holy Land*.
30. On Sanudo and Vesconte, see Harley and Woodward, *History of Cartography*.
31. Kent, *Cosimo de' Medici*, 431 n. 207; Larner, "Church and the *Quattrocento* Renaissance." Ptolemy's cartography has played a major role in Edgerton's scholarship; see his "From Mental Matrix to *Mappamundi* to Christian Empire." Edgerton also argued that Ptolemy's third projection was crucial for the invention of one-point perspective (*Renaissance Rediscovery*, 91–123).
32. Durand, *Vienna-Klosterneuburg Map Corpus*, 124–26. The group also made portable sundials, making it possible to establish location relative to coordinates.
33. Ibid., 343–45. *Pavimentum*, which means a stone or pounded earth floor and here evidently refers to a grid plane, is the word Alberti (*On Painting*, 20–21) used for his famous projected "checkerboard." Perhaps Alberti understood his method of establishing transversals and thus projecting a grid as his crucial contribution to painter's perspective.
34. Destombes, *Mappemondes*, 52.10, pl. 31.
35. Sterling, "Jan van Eyck avant 1432," 69–77.
36. Alberti (*On Painting*, 45) recommended that a wind god be placed in a corner of the history in order to specify the cause of the movement of draperies, which may also be regarded as a symbolic intrusion of cartographic into pictorial space.

37. See Edson, *Mapping Time and Space*, chaps. 2 and 3, for the classical and medieval history of these maps.
38. Harley and Woodward, *History of Cartography*, 334–35.
39. The single-footed, large-footed, reverse-footed, headless, dog-headed, long-eared, and generally monstrous races were usually associated with the East, and especially with India, in a tradition beginning in the early fourth century B.C.E. and given new impetus in the aftermath of the campaigns of Alexander the Great. Ignorance was compounded by fantasy in text and image through the Renaissance and given broad intercultural currency in the various Alexander romances. See the classic essays of Wittkower, "Marvels of the East" and "Marco Polo and the Pictorial Tradition."
40. In the Old Testament (Ezekiel 38–39), Gog is the foe of Jahweh; in the New Testament (Revelations 20:7), Gog and Magog are identified with Satan and overthrown and defeated at the end of time. Gog and Magog have similar apocalyptic meanings in the Koran (*Koran, with a Parallel Arabic Text*, 302, 329).
41. Nebenzahl, *Mapping the Silk Road and Beyond*, 26–27. For a full description, see Destombes, *Mappemondes*, 51.1 (p. 179). Besides the names of the sons of Noah and the four directions, inscriptions read in Asia: *l'arche Noe*, *le fenis*, *Ninive*, *Azareth*, *le lac meotide*, *arabe*; in Africa: *Athenes*; in Europe: *Maintz*.
42. A comparison made by Dhanens, *Hubert and Jan van Eyck*, 172–73.
43. Ptolemy, *Geography*, 25–26.
44. See Scafi, "Paradiso terrestre di Fra Mauro."
45. Meiss, "'Highlands' in the Lowlands"; Wohl, "Martin V and the Revival of the Arts in Rome." Masolino was a major beneficiary of this composition, expanding it to the scale of the spatial construction of Masaccio.
46. Alpers, *Art of Describing*, 119–68.
47. Gibson, *Mirror of the Earth*. Alpers, *Art of Describing*, observed that the first meta-optical landscapes—that is, landscapes shown as if measured but without a positioned viewer, comparable to a Mercator projection—are the world landscapes of Joachim Patenir.
48. Beach and Koch, *King of the World*, 84 n. 33; see also Koch, "Hierarchical Principles."

Chapter Four

1. Harley and Woodward, *History of Cartography*, 497.
2. For an introduction to Nicholas Cusanus, see Harries, *Infinity and Perspective*.
3. Jolly, "Jan van Eyck's Italian Pilgrimage," argued that van Eyck's *Annunciation* follows the pattern of this work, especially in the version by Lorenzo Monaco in the Salembeni Altarpiece in Santa Trinita, finished in these same years. The shrine of the Annunziata

was (and is) especially associated with pregnancy and childbirth, hence the interest of the Duke and Duchess of Burgundy (even if they were not the direct patrons of the altarpiece).

4. See most recently Holmes, *Fra Filippo Lippi*, 14–20. Holmes minimized the importance of the "German" Carmelite painter Fra Arrigo d'Arrigo, opting for a more purely Tuscan formation of Filippo's skills, following Ruda, "Flemish Painting"; Ruda, *Fra Filippo Lippi*, 126–33. Later in her argument, however, Holmes suggested that the "fascination with contemporary Flemish painting" (118) in Lippi's mature work owes to his earlier association with Fra Arrigo, and to my eye the interior in Filippo's *Tarquinia Madonna and Child* of 1437 is sufficient evidence of his admiring familiarity with slightly earlier Flemish painting, such as Robert Campin's *Salting Madonna* from around 1430.
5. Snyder, *Northern Renaissance Art*, 93.
6. Gombrich, "Light, Form, and Texture."
7. Vasari, *Vite*, 3:126.
8. Ponte, "Leon Battista Alberti e Genova," 69–70.
9. In his *Vita*, which was written in 1437 or 1438, Alberti spoke of letters he had written earlier to Toscanelli on astrology and prognostication (Watkins, "L. B. Alberti in the Mirror," 14). See also Flasch, "Nicolò Cusano e Leon Battista Alberti." The German Cusanus, born in 1401, was educated in Padua between 1417 and 1423, while the young Alberti studied in Padua from 1415 to 1418. Flasch concluded that although there is no documentary evidence of contact between them, the two moved and corresponded in the same humanist and ecclesiastical circles, and close contact is likely in their later years. Alberti died in 1472; Cusanus died and was buried in Rome in 1464.
10. See Morselli, "Proportions of Ghiberti's *St. Stephen*."
11. Baxandall, *Words for Pictures*, 31.
12. On Alberti's benefices (a second was added in 1449), see Boschetto, "Leon Battista Alberti e Firenze."
13. Mancini, *Vita*, 84–87. Kelly, *Leon Battista Alberti*, 5, accepted Mancini's arguments for Alberti's northern travels, noting, however, that Grayson expressed doubts in "Alberti, Leon Battista."
14. On Albergati, see *Dizionario biografico degli italiani*, 1:619–21.
15. Mancini, *Vita*, 60.
16. Ibid.
17. Hunter, "Who Is Jan van Eyck's 'Cardinal Niccolò Albergati'?," reviewed the evidence and the scholarship and concluded that the sitter was an unknown secular person.
18. Panofsky, *Early Netherlandish Painting*, 189–90, 199–201.
19. It has also been suggested that van Eyck attended the Council of Ferrara or that he was

associated with René d'Anjou, twice prisoner of Philip the Good, who returned to rule Naples in 1438 and was overthrown by Alfonso of Aragon; see Harbison, *Jan van Eyck*, 165–67.

20. Alberti, *On the Art of Building in Ten Books*, 54, 88, 179, 172.
21. Mancini, *Vita*, 11.
22. Ponte, "Leon Battista Alberti e Genova," 67.
23. Baxandall, *Giotto and the Orators*, 107, 166; Bialostocki, "Man and Mirror in Painting."
24. Alberti, *On Painting*, 19.
25. Kemp, *Science of Art*, 204.
26. Fubini and Gallorini, "Autobiografia di Leon Battista Alberti." I have also used Watkins, "L. B. Alberti in the Mirror," 11.
27. Ibid. This seafarer's advice is very old and was given broad symbolic currency by Jesus himself. See Matthew 16:2–3: "When it is evening, ye say, It will be fair weather, for the sky is red. And in the morning, It will be foul weather today: for the sky is red and lowring."
28. Pliny the Elder, *Natural History*, 35.xxxvi.95–96.
29. *Heures de Turin*, pls. 37 and 30.
30. Alberti, *On Painting*, 11.
31. Ibid., 19.
32. Ibid.
33. Alberti, *On Painting*, 18.
34. See chap. 2, n. 66.
35. For a review of the attribution, dating, and interpretation of this little dome, see Camerota, *Nel Segno di Masaccio*, 247.
36. See chap. 3, n. 32.
37. Vasari, *Vite*, 3:301–7.
38. Roy, "Van Eyck's Technique"; R. White, "Van Eyck's Technique"; Billinge, "Examining Jan van Eyck's Underdrawings."
39. See Nuttall, *From Flanders to Florence*, 161–70; Dunkerton, "North and South."
40. Meiss, "Light as Form and Symbol."
41. Snyder, *Northern Renaissance Art*, 100.
42. See chap. 2, n. 70.
43. Schwarz, "Mirror in Art."
44. Olson, *Florentine Tondo*, 97.
45. Ibid., 97 n. 79.
46. Summers, *Judgment of Sense*, 153–64.
47. Billinge, Verougstraete, and Van Schoutte, "The *Saint Barbara*." See also Gifford, "Van Eyck's Washington *Annunciation*."

48. Alberti, *On Painting*, 46–47.

49. Ibid., 11, 47.

50. Panofsky, *Early Netherlandish Painting*, 1:228–30.

51. Olson, *Florentine Tondo*, 95–101.

52. Edgerton, *Heritage of Giotto's Geometry*, 88–107; Edgerton and Steinberg, "How Shall This Be?"

53. Bartsch, "Philosopher as Narcissus," 72–73, citing McCarty, "Shape of the Mirror," 168–69.

54. Seneca, *Quaestiones naturales*, I.15.7.

55. Ibid., I.17.4.

56. Pliny the Elder, *Natural History*, 35.xxxvi.62–63; Alberti, *On Painting*, 25–26.

57. Alberti, *On Painting*, 26.

58. For a review of the extensive literature on the myth of Narcissus, see Bartsch, "Philosopher as Narcissus." Baskins, "Echoing Narcissus," considers Alberti's allusion in relation to "male parthenogenesis" and gender ambiguity. See also Barbieri, *Inventore della pittura*.

59. Quintilian, *Institutes*, 10.2.7.

60. Alberti, *On Painting*, 26.

61. According to Stoichita, *Short History of the Shadow*, 29–41, citing Jean Piaget and Jacques Lacan, the separation of shadow and reflection—the two are often used interchangeably in ancient and medieval texts—marks a passage from substitutive images to images understood as the basis of intersubjectivity.

62. Alberti, *On Painting*, 62 n. 11.

63. Philostratus the Elder, I.23.

64. Grafton, *Leon Battista Alberti*, 6, 278. Alberti may have studied Greek early on, and he cited many carefully read Greek authors when he wrote on architecture, but he was grateful for the appearance of Latin translations.

65. Baxandall, *Giotto and the Orators*, 100–103.

66. Watkins, "L. B. Alberti in the Mirror," 11. If Alberti could not have been in Florence before October 1428 (when he was in his mid-twenties), then according to this account he might have remembered his friends in Florence and discoursed with his friends in Venice in late 1429 or 1430.

67. Quintilian, *Institutes*, XII.I.1: "The orator, then, whom I am concerned to form, shall be the orator as defined by Marcus Cato, 'a good man, skilled in speaking.'"

68. Vasari, *Vite*, 3:289. See also introduction, n. 21.

69. Watkins, "L. B. Alberti's Emblem," 256–57; Schneider, "Leon Battista Alberti"; Woods-Marsden, *Renaissance Self-Portraiture*, 71–77.

70. Ovid, *Metamorphoses*, 3:335–510.

71. Ficino, *Commentary*, 140–41. Ficino cites Orpheus. See also Plotinus, *Ennead*, I.6.9.

72. Watkins, "L. B. Alberti in the Mirror," 26–27.

73. Freud, *Leonardo da Vinci*, 55. Freud's essay "On Narcissism: An Introduction" appeared in 1914. As an example of the generalizing cultural critical phrase, we may consider Lasch, *Culture of Narcissism*.

74. Kemp, "From Mimesis to Fantasia."

75. Summers, *Michelangelo*, 56–59.

76. Nuttall, *From Flanders to Florence*, 53–60.

77. Nicholas of Cusa, *De visione dei*, VIII.30 (p. 249), cited in Vesely, *Architecture in the Age of Divided Representation*, 157, together with a discussion of the friendship of Cusanus, Toscanelli, and Alberti. On Cusanus and Rogier van der Weyden, see Panofsky, "Facies illa Rogeri Maximi Pictoris."

78. According to Lane, "Patron and the Pirate," Angelo Tani (shown on the outside with his wife) commissioned the painting, finished by 1467, and the unscrupulous Portinari not only took Tani's position in the Medici bank in Bruges but also took over his painting and had his own portrait placed on the saved figure in Michael's scales; he is reflected as if facing the blessed rising.

79. Nicholas of Cusa, *De visione dei*, IX.35 (p. 250), X.40 (p. 253).

Afterword

1. Plato, *Republic*, 509d.
2. Vitruvius, *De architectura*, V.3.6–8.
3. Summers, *Judgment of Sense*, 1–31.
4. See Wood, "Perspective," for a summary of perspective's critical fortunes; for a detailed history, see Jay, *Downcast Eyes*.
5. Alberti, *On Painting*, 19.
6. Quoted in Guillén, "On the Concept and Metaphor of Perspective," 299.
7. Newton, *Opticks*, 370.
8. Capella, *Martianus Capella*, vol. 2, sec. 361, 118–19.
9. Heidegger, "The Origin of the Work of Art," in *Basic Writings*, 154.
10. Summers, "Cogito Embodied."
11. Leibniz, *Monadology*, in *Oeuvres*, 2:471–72. As noted by Guillén ("On the Concept and Metaphor of Perspective," 321 n. 75), Leibniz used this example several times; see also *Theodicy*, 357 (*Oeuvres*, 2:336) where he discusses the different mathematical forms by which a circle may be represented.
12. Aristotle (*De anima*, 433b19–21) argued that "the instrument by which the appetite (*orexis*) causes movement belongs already to the physical sphere; so it must be considered among the functions common to body and soul." His explanation is hardly better than Descartes'. The meeting of the two is where a beginning and an end coincide,

like a ball-and-socket joint. Appetite, he said, entails the capacity for self-movement, which in turn is impossible without imagination (*phantasia*).

13. Leibniz, *Theodicy*, in *Oeuvres*, 2:472.
14. Augustine, *City of God*, XI.18. See also Summers, "Contrapposto," 349–50.
15. Kant, *Critique of Pure Reason*, 155–92, on the "Transcendental Aesthetic."
16. Heidegger, *Kant and the Problem of Metaphysics*, 221.
17. The reference is to Jay, *Downcast Eyes*.
18. Antliff, *Inventing Bergson*.
19. Heidegger, *Kant and the Problem of Metaphysics*, 247. Bergson, Wilhelm Dilthey, and Georg Simmel are listed and linked in their "attempt to obtain a more searching and a more intuitive understanding of the 'liveliness' of life by determining its temporal character."
20. Leibniz, *Monadology*, in *Oeuvres*, 2:17, 64–74.
21. Most significantly in art-historical literature by Panofsky, as in his *Renaissance and Renascences in Western Art*, 108: "In the Italian Renaissance the classical past began to be looked upon from a fixed distance, quite comparable to the 'distance between the eye and the object' in that most characteristic invention of this very Renaissance, focused perspective . . . [T]his distance prohibited direct contact—owing to the interposition of an ideal 'projection plane'—but permitted a total and rationalized view." Paradoxically, Panofsky considered historical distance necessary for the longing that might yield a true renaissance. For a measured discussion of Panofsky's changing attitude toward historical distance and objectivity, see Holly, *Panofsky and the Foundations of Art History*, 188–93.
22. Stafford, *Body Criticism*, 464–77.

BIBLIOGRAPHY

Ackerman, James S. "Alberti's Light." In *Studies in Late Medieval and Renaissance Painting in Honor of Millard Meiss*, edited by Irving Lavin and John Plummer, 1–27. New York: New York University Press, 1977.

Alberti, Leon Battista. *Momus*. Translated by Sarah Knight. Cambridge: Harvard University Press, 2003.

———. *On Painting and on Sculpture: The Latin Texts of "De pictura" and "De statua."* Edited by Cecil Grayson. London: Phaidon, 1972.

———. *On the Art of Building in Ten Books*. Translated by Joseph Rykwert, Neil Leach, and Robert Tavernor. Cambridge: MIT Press, 1988.

———. *Opere volgari*. Vol. 3. Edited by Cecil Grayson. Bari: Laterza, 1973.

Albertus Magnus. *Opera omnia ad fidem codicum manuscriptorum edenda, apparatu critico, notis, prolegomenis, indicibus instruenda curavit Institutum Alberti Magni Coloniense*. Westphalia: Aschendorff, 1951–99.

Alhazen. *Opticae thesaurus: Alhazeni Arabis libri septem, nuncprimum editi: Eiusdem liber De Crepusculis et nubium ascensionibus: Item Vitellonis Thuringopoloni libri X*. Edited by David C. Lindberg. New York: Johnson, 1972.

———. *The Optics of Ibn al-Haytham: Books I–III: On Direct Vision*. 2 vols. Translated and introduction by A. I. Sabra. London: Warburg Institute, 1989.

Allen, Michael J. B. *Nuptial Arithmetic: Marsilio Ficino's Commentary on the Fatal Number in Book VIII of Plato's "Republic."* Berkeley: University of California Press, 1994.

Alpers, Svetlana. *The Art of Describing: Dutch Art in the Seventeenth Century*. Chicago: University of Chicago Press, 1983.

Ames-Lewis, Francis. *Drawing in Early Renaissance Italy*. New Haven: Yale University Press, 2000.

Antliff, Mark. *Inventing Bergson: Cultural Politics and the Parisian Avant-Garde*. Princeton: Princeton University Press, 1993.

Anzulewicz, Henryk. *De forma resultante in speculo: Die theologische Relevanz des Bildbegriffs un des Spiegelbildmodells in den frühwerken des Albertus Magnus*. Münster: Aschendorff, 1999.

Bacon, Roger. *The "Opus Majus" of Roger Bacon*. 2 vols. Edited by John Henry Bridges. Frankfurt am Main: Minerva, 1964.

Barasch, Moshe. *Light and Color in the Italian Renaissance Theory of Art*. New York: New York University Press, 1938.

Barbieri, Giuseppe. *L'inventore della pittura: Leon Battista Alberti e il mito del Narciso.* Vicenza: Terra Ferma, 2000.

Bartsch, Shadi. "The Philosopher as Narcissus: Vision, Sexuality and Self-Knowledge in Classical Antiquity." In *Visuality before and beyond the Renaissance*, edited by Robert S. Nelson, 70–97. Cambridge: Cambridge University Press.

Baskins, Cristelle L. "Echoing Narcissus in Alberti's *Della Pittura*." *Oxford Art Journal* 16 (1993): 25–33.

Baxandall, Michael. *Giotto and the Orators: Humanist Observers of Painting in Italy and the Discovery of Pictorial Composition.* Oxford: Oxford University Press, 1971.

———. *Painting and Experience in Fifteenth-Century Italy: A Primer in the Social History of Pictorial Style.* Oxford: Oxford University Press, 1988.

———. *Words for Pictures: Seven Papers on Renaissance Art and Criticism.* New Haven: Yale University Press, 2003.

Beach, Milo Cleveland, and Ebba Koch. *King of the World: The "Padshahnama," an Imperial Mughal Manuscript from the Royal Library, Windsor Castle.* London: Azimuth, 1997.

Behl, Benoy K. *The Ajanta Caves: Ancient Paintings of Buddhist India.* New York: Thames and Hudson, 2001.

Bergson, Henri. *Creative Evolution.* Edited by Pete A. Y. Gunter. New York: Barnes and Noble, 2005.

Bialostocki, Jan. "Man and Mirror in Painting: Reality and Transience." In *Studies in Late Medieval and Renaissance Painting in Honor of Millard Meiss*, edited by Irving Lavin and John Plummer, 61–82. New York: New York University Press, 1977.

Billinge, Rachel. "Examining Jan van Eyck's Underdrawings." In *Investigating Jan van Eyck*, edited by Susan Foister, Sue Jones, and Delphine Cool, 83–96. Turnhout: Brepols, 2000.

Billinge, Rachel, Hélène Verougstraete, and Roger Van Schoutte. "The *Saint Barbara*." In *Investigating Jan van Eyck*, edited by Susan Foister, Sue Jones, and Delphine Cool, 41–47. Turnhout: Brepols, 2000.

Borchert, Till-Holger. *The Age of Van Eyck: The Mediterranean World and Early Netherlandish Painting, 1430–1530.* London: Thames and Hudson, 2002.

Borsi, Franco, and Stefano Borsi. *Paolo Uccello.* Milan: Leonaro, 1992.

Boschetto, Luca. "Leon Battista Alberti e Firenze." In *Leon Battista Alberti e il Quattrocento: Studi in onore di Cecil Grayson e Ernst Gombrich: Atti del convegno internazionale, Mantova, 29-31 ottobre 1998*, edited by Luca Chiavoni, Gianfranco Ferlisi, Maria Vittoria Grassi, 435–50. Florence: Olschki, 2001.

Bourassin, Emmanuel. *Philippe le Bon: Le grand lion des Flandres.* Paris: Tallander, 1983.

Brown, A. L. "Three and Scene-Painting Sophocles." *Proceedings of the Cambridge Philological Society* 210 (n.s. 30) (1984): 1–17.

Bruno, Vincent J. *Form and Color in Greek Painting.* New York: Norton, 1977.
Buettner, Brigitte. *Boccaccio's "Des cleres et nobles femmes": Systems of Signification in an Illuminated Manuscript.* Seattle: University of Washington Press, 1996.
Burton, Harry Edwin. "The Optics of Euclid." *Journal of the Optical Society of America* 35 (1945): 357–72.
Camerota, Filippo, ed. *Nel segno di Masaccio: L'invenzione della prospettiva.* Florence: Giunti, 2001.
Caparello, Adriana. *Senso e interiorità in Alberto Magno.* Rome: Poliglotta della Pontificia Università Gregoriana, 1993.
Capella, Martianus. *Martianus Capella and the Seven Liberal Arts.* Vol. 2, *The Marriage of Philology and Mercury.* Translated by William Harris Stahl and Richard Johnson with Evan Laurie Burge. New York: Columbia University Press, 1977.
Chiavoni, Luca, Gianfranco Ferlisi, and Maria Vittoria Grassi, eds. *Leon Battista Alberti e il Quattrocento: Studi in onore di Cecil Grayson e Ernst Gombrich: Atti del convegno internazionale, Mantova, 29–31 ottobre 1998.* Florence: Olschki, 2001.
Christiansen, Keith, ed. *From Filippo Lippi to Piero della Francesca: Fra Carnevale and the Making of a Renaissance.* New Haven: Yale University Press, 2005.
Coremans, Paul B. *L'agneau mystique au laboratoire: Examen et traitement.* Antwerp: De Sikkel, 1953.
Crary, Jonathan. *Suspensions of Perception: Attention, Spectacle, and Modern Culture.* Cambridge: MIT Press, 1999.
———. *Techniques of the Observer: On Vision and Modernity in the Nineteenth Century.* Cambridge: MIT Press, 1991.
Cyriac of Ancona. *Later Travels.* Edited and translated by Edward W. Bodnar with Clive Foss. Cambridge: Harvard University Press, 2003.
Dante Alighieri. *Convivio.* Vol. 3 of *Le opere di Dante Alighieri.* 3 vols. in 8. Edited by Franca Brambilla Ageno. Florence: Casa Editrice le Lettere, 1995.
Degl'Innocenti, Giovanni. "Il dimensionamento della tavoletta del primo esperimento prospettico brunelleschiano." In *Filippo Brunelleschi: La sua opera e il suo tempo: Atti del Convegno internazionale di studi, Firenze, 16–22 ottobre 1977,* 2:561–70. Florence: Centro Di, 1980.
Demosthenes. *Against Meidias: Oration 21.* Edited with introduction, translation, and commentary by Douglas M. MacDowell. New York: Oxford University Press, 1990.
Destombes, Marcel. *Mappemondes A.D. 1200–1500.* Amsterdam: Israel, 1964.
Dhanens, Elisabeth. *The Ghent Altarpiece.* New York: Viking, 1973.
———. *Hubert and Jan van Eyck.* New York: Alpine Fine Arts, 1980.
Dizionario biografico degli italiani. Rome: Istituto della Enciclopedia Italiana, 1960–2005.
Dunkerton, Jill. "North and South: Painting Techniques in Venice." In *Renaissance*

Venice and the North: Crosscurrents in the Time of Bellini, Dürer and Titian, edited by Bernard Aikema and Beverly Louise Brown, 93–103. New York: Rizzoli, 1999.

Durand, Dana Bennett. *The Vienna-Klosterneuburg Map Corpus of the Fifteenth Century: A Study in the Transition from Medieval to Modern Science*. Leiden: Brill, 1952.

Edgerton, Samuel Y., Jr. "From Mental Matrix to *Mappamundi* to Christian Empire: The Heritage of Ptolemaic Cartography in the Renaissance." In *Art and Cartography: Six Historical Essays*, edited by David Woodward, 10–50. Chicago: University of Chicago Press, 1987.

———. *The Heritage of Giotto's Geometry: Art and Science on the Eve of the Scientific Revolution*. Ithaca: Cornell University Press, 1991.

———. *The Renaissance Rediscovery of Linear Perspective*. New York: Basic Books, 1975.

Edgerton, Samuel Y., Jr., and Leo Steinberg. "How Shall This Be?: Reflections on Filippo Lippi's *Annunciation* in London, Parts I and II." *Artibus et Historiae* 16 (1987): 25–55.

Edson, Evelyn. *Mapping Time and Space: How Medieval Mapmakers Viewed Their World*. London: British Library, 1997.

Elkins, James. "On the *Arnolfini Portrait* and the Lucca *Madonna*: Did Jan van Eyck Have a Perspective System?" *Art Bulletin* 73 (1991): 53–62.

Erlande-Brandenburg, Alain, Regine Pernoud, Jean Gimpel, and Roland Bechmann. *Carnet de Villard de Honnecourt: D'après le manuscrit conservé à la Bibliothèque Nationale de Paris (no. 19093)*. Paris: Stock, 1986.

Farago, Claire J. *Leonardo da Vinci's "Paragone": A Critical Interpretation with a new Edition of the Text in the Codex Urbinas*. Leiden: Brill, 1992.

———. "Leonardo's Color and Chiaroscuro Reconsidered: The Visual Force of Painted Images." *Art Bulletin* 73 (1991): 63–88.

Federici-Vescovini, Graziella. "L'inserimento della 'perspectiva' tra le arti del quadrivio." In *Arts libéraux et philosophie au Moyen Age: Actes*, 969–74. Montreal: Institut d'Études Médiévales, 1969.

———. "Il problema delle fonti ottiche medievali del *Commentario terzo* di Lorenzo Ghiberti." In *Lorenzo Ghiberti nel suo tempo: Atti del Convegno internazionale di studi, Firenze, 18–21 ottobre 1978*, 349–87. Florence: Olschki, 1980.

———. "La prospettiva del Brunelleschi: Alhazen e Biagio Pelacani a Firenze." In *Filippo Brunelleschi: La sua opera e il suo tempo: Atti del Convegno internazionale di studi, Firenze, 16–22 ottobre 1977*, 1:333–48. Florence: Centro Di, 1980.

———. "Le questioni di 'Perspectiva' di Biagio Pelacani da Parma." *Rinascimento*, 2d ser., 1 (1961): 163–243.

Fehrenbach, Frank. "Der oszillierende Blick: *Sfumato* und die Optik des späten Leonardo." *Zeitschrift für Kunstgeschichte* 65 (2002): 522–44.

Ficino, Marsilio. *Commentary on Plato's Symposium on Love*. Translated by Sears Jayne. Woodstock, Conn.: Spring, 1985.

Filarete. *Treatise on Architecture: Being the Treatise by Antonio di Piero Averlino, Known as Filarete*. Edited and translated by John Richard Spencer. New Haven: Yale University Press, 1965.

Fiorani, Francesca. *The Marvel of Maps: Art, Cartography, and Politics in Renaissance Italy*. New Haven: Yale University Press, 2005.

Flasch, Kurt. "Nicolò Cusano e Leon Battista Alberti." In *Leon Battista Alberti e il Quattrocento: Studi in onore di Cecil Grayson e Ernst Gombrich: Atti del convegno internazionale, Mantova, 29-31 ottobre 1998*, edited by Luca Chiavoni, Gianfranco Ferlisi, Maria Vittoria Grassi, 371–80. Florence: Olschki, 2001.

Foister, Susan, Sue Jones, and Delphine Cool, eds. *Investigating Jan van Eyck*. Turnhout: Brepols, 2000.

Freud, Sigmund. *Leonardo da Vinci and a Memory of His Childhood*. New York: Norton, 1964.

Fubini, Riccardo, and Anna Menci Gallorini. "L'autobiografia de Leon Battista Alberti: Studio e edizione." *Rinascimento*, 2d ser., 12 (1972): 21–78.

Fumagalli, Giuseppina. *Leonardo: Omo senza lettere*. Florence: Sansoni, 1952.

Gage, John. *Color and Meaning: Art, Science, and Symbolism*. Berkeley: University of California Press, 1999.

Galen. *Galen on the Usefulness of the Parts of the Body: De usu partium*. Translated and introduction by Margaret Tallmadge May. Ithaca: Cornell University Press, 1968.

Garin, Eugenio. *La cultura filosofica del Rinascimento italiano: Ricerche e documenti*. 2d ed. Florence: Sansoni, 1979.

Gauricus, Pomponius. *De sculptura*. Edited by Paolo Cutolo. Naples: Scientifiche Italiane, 1999.

Gherardi da Prato, Giovanni. *Il paradiso degli Alberti*. Edited by Antonio Lanza. Rome: Salerno, 1975.

Ghiberti, Lorenzo. *Lorenzo Ghiberti: I Commentarii: Biblioteca Nazionale di Firenze, II.I.333*. Edited and introduction by Lorenzo Bartoli. Florence: Giunti, 1998.

Gibson, Walter. *Mirror of the Earth: The World Landscape in Sixteenth-Century Flemish Painting*. Princeton: Princeton University Press, 1989.

Gifford, E. Melanie. "Van Eyck's Washington *Annunciation*: Technical Evidence for Iconographic Development." *Art Bulletin* 81 (1999): 108–16.

Goffen, Rona. *Masaccio's "Trinity."* Cambridge: Cambridge University Press, 1998.

Goldthwaite, Richard A. *The Building of Renaissance Florence: An Economic and Social History*. Baltimore: Johns Hopkins University Press, 1980.

Gombrich, Ernst Hans. *Art and Illusion: A Study in the Psychology of Pictorial Representation*. Princeton: Princeton University Press, 2000.

———. "From the Revival of Letters to the Reform of the Arts: Niccolò Niccoli and

Filippo Brunelleschi." In *The Essential Gombrich: Selected Writings on Art and Culture*, edited by Richard Woodfield, 411–35. London: Phaidon, 1996.
———. "The Heritage of Apelles." In *The Heritage of Apelles: Studies in the Art of the Renaissance*, 3–18. Ithaca: Cornell University Press, 1976.
———. "Light, Form, and Texture in Fifteenth-Century Painting North and South of the Alps." In *The Heritage of Apelles: Studies in the Art of the Renaissance*, 19–35. Ithaca: Cornell University Press, 1976.
———. *The Preference for the Primitive: Episodes in the History of Western Art and Taste.* London: Phaidon, 2002.
———. "The 'What' and the 'How': Perspective Representation and the Phenomenal World." In *Logic and Art: Essays in Honor of Nelson Goodman*, edited by Richard Rudner and Israel Scheffler, 129–49. Indianapolis: Bobbs-Merrill, 1972.
Goodman, Nelson. *Languages of Art: An Approach to the Theory of Symbols.* 2d ed. Indianapolis: Hackett, 1976.
Grafton, Anthony. *Leon Battista Alberti: Master Builder of the Renaissance.* New York: Hill and Wang, 2000.
Grayson, Cecil. "Alberti, Leon Battista." In *Dizionario biografico degli italiani*, 1:702–9. Rome: Istituto della Enciclopedia Italiana, 1960–2005.
Greenstein, Jack M. "On Alberti's 'Sign': Vision and Composition in Quattrocento Painting." *Art Bulletin* 79 (1997): 669–98.
Guillén, Claudio. "On the Concept and Metaphor of Perspective." In *Literature as System: Essays toward the Theory of Literary History*, 283–371. Princeton: Princeton University Press, 1971.
Hall, Edwin. *The Arnolfini Betrothal: The Medieval Marriage and the Enigma of van Eyck's Double Portrait.* Berkeley: University of California Press, 1994.
Harbison, Craig. *Jan van Eyck: The Play of Realism.* London: Reaktion, 1991.
Harley, J. B., and David Woodward. *History of Cartography.* Vol. 1, *Cartography in Prehistoric, Ancient, and Medieval Europe.* Chicago: University of Chicago Press, 1987.
Harries, Karsten. *Infinity and Perspective.* Cambridge: MIT Press, 2001.
Heidegger, Martin. *Basic Writings from "Being and Time" (1927) to "The Task of Thinking" (1964).* Edited by David Farrell Krell. New York: Harper and Row, 1977.
———. *Kant and the Problem of Metaphysics.* Translated by James S. Churchill. Bloomington: Indiana University Press, 1962.
Heilbron, J. L. *The Sun in the Church: Cathedrals as Solar Observatories.* Cambridge: Harvard University Press, 2001.
Heures de Turin: Quarante-cinq feuillets à peintures provenant des Très belles heures de Jean de France, duc de Berry: Reproduction en phototype d'après les Originaux de la Biblioteca Nazionale de Turin et du Musée du Louvre. Introduction by Albert Chatelet. Turin: Bottega d'Erasmo, 1967.

Hills, Paul. *The Light of Early Italian Painting*. New Haven: Yale University Press, 1987.
Hobey-Hamsher, Catherine. "Agatharcos." In *The Dictionary of Art*, edited by Jane S. Turner, 1:450. New York: Grove, 1996.
———. "Apollodoros." In *The Dictionary of Art*, edited by Jane S. Turner, 2:227. New York: Grove, 1996.
Hockney, David. *Secret Knowledge: Rediscovering the Lost Techniques of the Old Masters*. London: Thames and Hudson, 2001.
Hollanda, Francisco de. *Four Dialogues on Painting*. Translated by Audrey F. G. Bell. Westport, Conn.: Hyperion, 1979.
Holly, Michael Ann. *Panofsky and the Foundations of Art History*. Ithaca: Cornell University Press, 1984.
Holmes, Megan. *Fra Filippo Lippi: The Carmelite Painter*. New Haven: Yale University Press, 1999.
Hornblower, Simon, and Anthony Spawforth, eds. *The Oxford Classical Dictionary*. 3d ed. Oxford: Oxford University Press, 1996.
Hunter, John. "Who Is Jan van Eyck's 'Cardinal Niccolò Albergati'?" *Art Bulletin* 75 (1993): 207–18.
Hyman, Isabelle, comp. *Brunelleschi in Perspective*. Englewood Cliffs, N.J.: Prentice-Hall, 1974.
Jay, Martin. *Downcast Eyes: The Denigration of Vision in Twentieth-Century French Thought*. Berkeley: University of California Press, 1993.
Jolly, Penny Howell. "Jan van Eyck's Italian Pilgrimage: A Miraculous Florentine *Annunciation* and the Ghent Altarpiece." *Zeitschrift fuer Kunstgeschichte* 61 (1998): 369–94.
Kant, Immanuel. *Critique of Pure Reason*. Edited and translated by Paul Guyer and Allen W. Wood. Cambridge: Cambridge University Press, 1998.
Kaufmann, Thomas da Costa. "The Perspective of Shadows: The History of the Theory of Shadow Projection." *Journal of the Warburg and Courtauld Institutes* 38 (1975): 258–87.
Kelly, Joan. *Leon Battista Alberti: Universal Man of the Early Renaissance*. Chicago: University of Chicago Press, 1969.
Kemp, Martin. "From Mimesis to Fantasia: The Quattrocento Vocabulary of Creation, Inspiration, and Genius in the Visual Arts," *Viator* 8 (1977): 128–49.
———. "Leonardo and the Visual Pyramid." *Journal of the Warburg and Courtauld Institutes* 40 (1977): 128–49.
———. "The Mean and Measure of All Things." In *Circa 1492: Art in the Age of Exploration*, edited by Jay A. Levenson, 95–111. New Haven: Yale University Press, 1991.

———. *The Science of Art: Optical Themes in Western Art from Brunelleschi to Seurat.* New Haven: Yale University Press, 1990.

Kent, Dale. *Cosimo de' Medici and the Florentine Renaissance: The Patron's Oeuvre.* New Haven: Yale University Press, 2000.

Keuls, Eva C. *Plato and Greek Painting.* Leiden: Brill, 1978.

———. "*Skiagraphia* Once Again." *American Journal of Archeology* 79 (1975): 1–16.

Klein, Robert. "Pomponius Gauricus on Perspective." In *Form and Meaning: Essays on the Renaissance and Modern Art*, 102–28. Princeton: Princeton University Press, 1979.

Koch, Ebba. "The Hierarchical Principles of Shah-Jahani Painting." In *King of the World: The "Padshahnama," an Imperial Mughal Manuscript from the Royal Library, Windsor Castle*, edited by Milo Cleveland Beach and Ebba Koch, 130–43. London: Azimuth, 1997.

The Koran, with a Parallel Arabic Text. Translated by N. J. Dawood. London: Penguin, 2000.

Krautheimer, Richard, and Trudy Krautheimer-Hess. *Lorenzo Ghiberti.* Princeton: Princeton University Press, 1970.

Kretzmann, Norman, Anthony Kenny, Jan Pinborg, and Eleonore Stump, eds. *The Cambridge History of Later Medieval Philosophy: From the Rediscovery of Aristotle to the Disintegration of Scholasticism, 1100–1600.* Cambridge: Cambridge University Press, 1988.

Landino, Cristoforo. *Comento sopra la Comedia.* 4 vols. Edited by Paolo Procaccioli. Rome: Salerno, 2001.

Lane, Barbara. "The Patron and the Pirate: The Mystery of Memling's Gdansk *Last Judgment.*" *Art Bulletin* 73 (1991): 623–40.

Lang, Susanne. "Brunelleschi's Panels." In *La prospettiva rinascimentale: Codificazioni e transgressioni*, edited by Marisa Dalai Emiliani, 1:63–72. Florence: Centro Di, 1980.

Larner, John. "The Church and the *Quattrocento* Renaissance in Geography." *Renaissance Studies* 12 (1998): 26–39.

Lasch, Christopher. *The Culture of Narcissism: American Life in an Age of Diminishing Expectations.* New York: Norton, 1978.

Lee, Rensselaer W. "*Ut pictura poesis*: The Humanistic Theory of Painting." *Art Bulletin* 22 (1940): 197–269.

Leibniz, Gottfried Wilhelm. *Oeuvres de Leibniz.* 2 vols. Edited by M. A. Jacques. Paris: Charpentier, 1842.

Leopold, Antony. *How to Recover the Holy Land: The Crusade Proposals of the Late Thirteenth and Fourteenth Centuries.* Aldershot: Ashgate, 2000.

Levenson, Jay A., ed. *Circa 1492: Art in the Age of Exploration.* New Haven: Yale University Press, 1991.

Lindberg, David C. "Lenses and Eyeglasses." In *Dictionary of the Middle Ages*, edited by Joseph R. Strayer, 7:538–41. New York: Scribner, 1986.

———. *Roger Bacon and the Origins of "Perspectiva" in the Middle Ages: A Critical Edition and English Translation of Bacon's "Perspectiva" with Introduction and Notes.* New York: Oxford University Press, 1996.

———. *Theories of Vision from al-Kindi to Kepler*. Chicago: University of Chicago Press, 1976.

Lydakis, Stelios. *Ancient Greek Painting and Its Echoes in Later Art*. Los Angeles: Getty Museum, 2004.

Macrobius, Ambrosius Aurelius Theodosius. *Saturnalia: Apparatu critico instruxit, in Somnium Scipionis commentarios selecta varietate lectionis*. 2 vols. Edited by Iacobus Willis. Leipzig: Teubner, 1970.

Mancini, Girolamo. *Vita di Leon Battista Alberti*. 2d ed. Rome: Bardi, 1971.

Manetti, Antonio. *The Life of Brunelleschi by Antonio di Tuccio Manetti*. Edited by Howard Saalman. University Park: Pennsylvania State University Press, 1970.

———. *Vita di Filippo Brunelleschi: Preceduta da la Novella del Grasso*. Edited by Domenico de Robertis. Milan: Polifilo, 1976.

McCarty, Willard. "The Shape of the Mirror: Metaphorical Catoptrics in Classical Literature." *Arethusa* 22 (1989): 161–95.

Meiss, Millard. "'Highlands' in the Lowlands." *Gazette des Beaux Arts* 103 (1963): 273–314.

———. "Light as Form and Symbol in Some Fifteenth-Century Paintings." In *The Painter's Choice: Problems in the Interpretation of Renaissance Art*, 3–18. New York: Harper and Row, 1976.

Moreno, Paolo. *Apelles: The Alexander Mosaic*. London: Skira, 2001.

Morselli, Piero. "The Proportions of Ghiberti's *St. Stephen*: Vitruvius's *De architectura* and Alberti's *De statua*." *Art Bulletin* 60 (1978): 235–41.

Muecke, Frances "'Taught by Love': The Origin of Painting Again." *Art Bulletin* 81 (1999): 299–302.

Nebenzahl, Kenneth. *Mapping the Silk Road and Beyond: 2000 Years of Exploring the East*. London: Phaidon, 2004.

Nelson, Robert S. "To Say and to See: Ekphrasis and Vision in Byzantium." In *Visuality before and beyond the Renaissance: Seeing as Others Saw*, edited by Robert S. Nelson, 143–68. Cambridge: Cambridge University Press, 2000.

Newton, Isaac. *Opticks; or, A Treatise of the Reflections, Refractions, Inflections, and Colours of Light: Based on the 4th ed., London, 1730*. New York: Dover, 1979.

Nicholas of Cusa. *De visione dei*. In *Selected Spiritual Writings*, translated by H. Lawrence Bond, 233–89. New York: Paulist, 1997.

Nuttall, Paula. *From Flanders to Florence: The Impact of Netherlandish Painting, 1400–1500*. New Haven: Yale University Press, 2004.
Olson, Roberta J. M. *The Florentine Tondo*. Oxford: Oxford University Press, 2000.
Panofsky, Erwin. *Early Netherlandish Painting: Its Origin and Character*. 2 vols. Cambridge: Harvard University Press, 1953.
———. "Facies illa Rogeri Maximi Pictoris." In *Late Classical and Medieval Studies in Honor of Albert Mathias Friend, Jr.*, edited by Kurt Weitzmann, 392–400. Princeton: Princeton University Press, 1955.
———. *Perspective as Symbolic Form*. Translated by Christopher S. Wood. New York: Zone, 1991.
———. *Renaissance and Renascences in Western Art*. New York: Harper and Row, 1960.
Parronchi, Alessandro. *Studi sulla dolce prospettiva*. Milan: Martello, 1964.
Paviot, Jacques. "Burgundy and the Crusade." In *Crusading in the Fifteenth Century: Message and Import*, edited by Norman Housley, 70–80. Houndmills: Palgrave Macmillan, 2004.
———. *Les ducs de Bourgogne, la Croisade, et l'Orient: Fin XIVe siècle–XVe siècle*. Paris: Presses de l'Université de Paris-Sorbonne, 2003.
———. "La Mappemonde attribuée a Jan van Eyck par Facio: Une piece a retirer de son oeuvre." *Revue des archeologues et historiens de l'art de Louvain* 24 (1991): 57–62.
Peckham, John. *John Pecham and the Science of Optics: Perspectiva Communis*. Edited and introduction by David C. Lindberg. Madison: University of Wisconsin Press, 1970.
Pemberton, Elizabeth G. "A Note on *Skiagraphia*." *American Journal of Archeology* 80 (1976): 82–84.
Philip, Lotte Brand. *The Ghent Altarpiece and the Art of Jan van Eyck*. Princeton: Princeton University Press, 1971.
Planchart, Alejandro Enrique. "Guillaume Dufay." In *The New Grove Dictionary of Music and Musicians*, 2d ed., edited by Stanley Sadie, 7:647–64. New York: Grove, 2001.
Pollitt, Jerome Jordan. *The Ancient View of Greek Art: Criticism, History and Terminology*. New Haven: Yale University Press, 1974.
———. *The Art of Ancient Greece: Sources and Documents*. Cambridge: Cambridge University Press, 1990.
Ponte, Giovanni. "Leon Battista Alberti e Genova." In *Leon Battista Alberti e il Quattrocento: Studi in Onore di Cecil Grayson e Ernst Gombrich: Atti del Convegno Internazionale, Mantova 29–31 ottobre 1998*, edited by Luca Chiavoni, Gianfranco Ferlisi, and Maria Vittoria Grassi, 65–80. Florence: Olschki, 2001.
Praeger, Frank D., and Gustina Scaglia. *Brunelleschi: Studies of His Technology and Inventions*. Cambridge: MIT Press. 1970.

Ptolemy, Claudius. *The Geography*. Translated and edited by Edward Luther Stevenson. Mineola, N.Y.: Dover, 1991.

Raynaud, Dominique. "Perspectiva naturalis." In *Nel segno di Masaccio: L'invenzione di prospettiva*, edited by Filippo Camerota, 11–18. Florence: Giunti, 2001.

Richter, Gisela M. A. *Perspective in Greek and Roman Art*. London: Phaidon, 1970.

Robertson, Martin. *Greek Painting*. Geneva: Skira, 1979.

Roy, Ashok. "Van Eyck's Technique: The Myth and the Reality I." In *Investigating Jan van Eyck*, edited by Susan Foister, Sue Jones, and Delphine Cool, 97–101. Turnhout: Brepols, 2000.

Ruda, Jeffrey. "Flemish Painting and the Early Renaissance in Florence: Questions of Influence." *Zeitschrift für Kunstgeschichte* 47 (1984): 210–36.

———. *Fra Filippo Lippi: Life and Work with a Complete Catalogue*. New York: Abrams, 1993.

Russell, Donald Andrew, and Michael Winterbottom, eds. *Ancient Literary Criticism*. Oxford: Clarendon, 1972.

Rykwert, Joseph, and Anne Engel, eds. *Leon Battista Alberti*. Milan: Electa, 1994.

Scafi, Alessandro. "Il paradiso terrestre di Fra Mauro." *Storia dell'Arte* 93/94 (1998): 411–19.

Schneider, Laurie. "Leon Battista Alberti: Some Biographical Implications of the Winged Eye." *Art Bulletin* 72 (1990): 261–70.

Schoene, Wolfgang. *Über das Licht in der Malerei*. Berlin: Mann, 1954.

Schwarz, Heinrich. "The Mirror in Art." *Art Quarterly* 15 (1952): 96–118.

Sinn, Ulrich, and Irma Wehgartner. *Begegnungen mit der Antike: Zeugnisse aus vier Jahrtausenden mittlelmeerischer Kultur in Südflügel der Würzburger Residenz*. Würzburg: Ergon, 2001.

Smith, A. Mark. *Ptolemy's Theory of Visual Perception: An English Translation of the "Optics."* Philadelphia: American Philosophical Society, 1986.

Snyder, James. *Northern Renaissance Art: Painting, Sculpture, the Graphic Arts from 1350 to 1575*. New York: Abrams, 1985.

Stafford, Barbara Maria. *Body Criticism: Imaging the Unseen in Enlightenment Art and Medicine*. Cambridge: MIT Press, 1991.

Sterling, Charles. "Jan van Eyck avant 1432." *Revue de l'Art* 33 (1976): 7–82.

Stoichita, Victor I. *A Short History of the Shadow*. London: Reaktion, 1997.

Summers, David. "Cogito Embodied: Force and Counterforce in René Descartes's *Les passions du l'âme*." In *Representing the Passions: Histories, Bodies, Visions*, edited by Richard Meyer, 13–36. Los Angeles: Getty Research Institute, 2003.

———. "Contrapposto: Style and Meaning in Renaissance Art." *Art Bulletin* 59 (1977): 336–61.

———. "The Heritage of Agatharcus: On Naturalism and Theatre in European Painting." In *The Beholder: The Experience of Art in Early Modern Europe*, edited by Thomas Frangenberg and Robert Williams, 9–36. Burlington: Ashgate, 2006.

———. *The Judgment of Sense: Renaissance Naturalism and the Rise of Aesthetics*. Cambridge: Cambridge University Press, 1987.

———. *Michelangelo and the Language of Art*. Princeton: Princeton University Press, 1981.

———. "Pandora's Crown: On Wonder, Imitation, and Mechanism in Western Art." In *Wonders, Marvels, and Monsters in Early Modern Culture*, edited by Peter G. Platt, 45–75. Newark: University of Delaware Press, 1999.

———. *Real Spaces: World Art History and the Rise of Western Modernism*. London: Phaidon, 2003.

Takahashi, Ken'ichi. *The Medieval Latin Traditions of Euclid's "Catoptrica": A Critical Edition of "De speculis" with an Introduction, English Translation, and Commentary*. Fukuoka: Kyushu University Press, 1992.

Tanturli, G. "Rapporti del Brunelleschi con gli ambienti letterari fiorentini." In *Filippo Brunelleschi: La sua opera e il suo tempo: Atti del Convegno internazionale di studi, Firenze, 16–22 ottobre 1977*, 1:125–44. Florence: Centro Di, 1980.

Toscanelli, Paolo dal Pozzo. *Della prospettiva*. Edited by Alessandro Parronchi. Milan: Polifilo, 1991.

Trachtenberg, Marvin. *Dominion of the Eye: Urbanism, Art, and Power in Early Modern Florence*. Cambridge: Cambridge University Press, 1977.

Vasari, Giorgio. *Le vite de' più eccellenti pittori, scultori, e architettori nelle redazioni del 1550 e 1568*. Edited by Rosanna Bettarini and Paola Barocchi. 6 vols. in 11. Florence: Sansoni, 1967–87.

Vesely, Dalibor. *Architecture in the Age of Divided Representation: The Question of Creativity in the Shadow of Production*. Cambridge: MIT Press, 2004.

Voisin, Jean-Louis. "Exoriente sole (Suetone, Ner. 6). D'Alexandrie a la Domus Aurea." In *L'urbs: Espace urbain et histoire (Ier siècle av. J.-C.—IIIe siècle ap. J.-C.): Actes du colloque international organisé par le Centre national de la recherche scientifique et l'École française de Rome (Rome, 8–12 mai 1985)*, 509–43. Rome: École Française de Rome, 1987.

Waal, Hans van de. "The *Linea Summae Tenuitatis* of Apelles: Pliny's Phrase and Its Interpreters." *Zeitschrift für Aesthetik und allgemeine Kunstwissenschaft* 12 (1967): 5–32.

Watkins, Renée N. "L. B. Alberti in the Mirror: An Interpretation of the *Vita* with a New Translation." *Italian Quarterly* 30 (1989): 5–30.

———. "L. B. Alberti's Emblem, the Winged Eye, and His Name, Leo." *Mitteilungen des kunsthistorischen Institutes in Florenz* 9 (1960): 256–58.

White, John. *The Birth and Rebirth of Pictorial Space*. 3d ed. Cambridge: Harvard University Press, 1987.

White, Raymond. "Van Eyck's Technique: The Myth and the Reality II." In *Investigating Jan van Eyck*, edited by Susan Foister, Sue Jones, and Delphine Cool, 101–6. Turnhout: Brepols, 2000.

Wittkower, Rudolf. "Marco Polo and the Pictorial Tradition of the Marvels of the East." In *Allegory and the Migration of Symbols*, 75–92. London: Thames and Hudson, 1977.

———. "Marvels of the East: A Study of the History of Monsters." In *Allegory and the Migration of Symbols*, 45–74. London: Thames and Hudson, 1977.

Wohl, Helmut. "Martin V and the Revival of the Arts in Rome." In *Rome in the Renaissance: The City and the Myth: Papers of the Thirteenth Annual Conference of the Center for Medieval and Early Renaissance Studies*, edited by P. A. Ramsey, 171–83. Binghamton, N.Y.: Center for Medieval and Early Renaissance Studies, 1982.

Wood, Christopher S. "Perspective: An Overview." In *Encyclopedia of Aesthetics*, edited by Michael Kelly, 3:477–81. New York: Oxford University Press, 1998.

Woods-Marsden, Joanna. *Renaissance Self-Portraiture: The Visual Construction of Identity and the Social Status of the Artist*. New Haven: Yale University Press, 1998.

Woodward, David, ed. *Art and Cartography: Six Historical Essays*. Chicago: University of Chicago Press, 1987.

Wright, Craig. "Dufay's *Nuper rosarum flores*, King Solomon's Temple, and the Veneration of the Virgin." *Journal of the American Musicological Society* 47 (1994): 395–441.

INDEX

Note: Italic page numbers refer to illustrations.